# The Concept of Probability

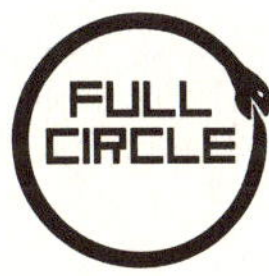


*Publications of the Archive of Scientific Philosophy*
*Hillman Library, University of Pittsburgh*

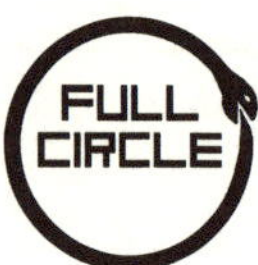

*Publications of the Archive of Scientific Philosophy*
*Hillman Library, University of Pittsburgh*

Volume 1.
*Frege's Lectures on Logic: Carnap's Student Notes, 1910–1914*, edited by Erich H. Reck and Steve Awodey

Volume 2.
*Carnap Brought Home: The View From Jena*, edited by Steve Awodey and Carsten Klein

Volume 3.
*The Concept of Probability in the Mathematical Representation of Reality, by Hans Reichenbach*, translated and edited by Frederick Eberhardt and Clark Glymour

Volume 4.
*Empiricism at the Crossroads: The Vienna Circle's Protocol-Sentence Debate*, by Thomas Uebel

# The Concept of Probability in the Mathematical Representation of Reality

Hans Reichenbach

*Translated and edited by*
Frederick Eberhardt and Clark Glymour

Open Court
Chicago and LaSalle, Illinois

**To order books from Open Court, call toll-free 1-800-815-2280, or visit our website at www.opencourtbooks.com.**

Open Court Publishing Company is a division of Carus Publishing Company.



First printing 2008



Printed and bound in the United States of America.

Designed by John Grandits.

Typeset in LaTeX by Dirk Schlimm.

The watermark image on the back cover comes from Reichenbach's original sketch for his probability machine (see p. 7). Reproduced courtesy of the Archives of Scientific Philosophy, Hillman Library, University of Pittsburgh (ASP/HR 025-04-01).

**Library of Congress Cataloging-in-Publication Data**

Reichenbach, Hans, 1891–1953. [Bedeutung der Wahrscheinlichkeit für die mathematische Darstellung der Wirklichkeit. English & German] The concept of probability in the mathematical representation of reality / Hans Reichenbach ; translated and edited by Frederick Eberhardt and Clark Glymour.

p. cm. – (Full circle ; v. 3)

English and German. Originally presented as the author's thesis (doctoral)–Friedrich-Alexander University Erlangen, 1915.

Includes bibliographical references (p. ) and index.

ISBN-13: 978-0-8126-9609-7 (trade paper : alk. paper)

ISBN-10: 0-8126-9609-3 (trade paper : alk. paper)

1. Kant, Immanuel, 1724–1804. Kritik der reinen Vernunft. 2. Knowledge, Theory of. 3. Causation. 4. Reason. 5. Logic, Symbolic and mathematical. 6. Probabilities. 7. Science–Philosophy. 8. Mathematics–Philosophy. I. Eberhardt, Frederick, 1978- II. Glymour, Clark N. III. Title.

B2779.R45 2007

121–dc22 2007021145

# Contents

# The Concept of Probability

# Acknowledgements

Our special thanks goes to Flavia Padovani who helped with this volume in many ways. We are indebted to her for valuable comments on a draft of the translation, for pointing out many mistakes, and for suggesting several improvements in the translation of particular terms. Flavia also pointed us to many sources that provided more background on Reichenbach's thesis work and its context. We are grateful to Beryl Schlossman for help with a passage from Cournot, to Bilge Mutlu for reconstructing Reichenbach's graphs, to Brigitta Arden for all the help with Reichenbach's unpublished notes at the Reichenbach Archives at the University of Pittsburgh, and to Dirk Schlimm for typesetting the entire manuscript. Finally, our thanks to Steve Awodey for the very idea of this project. None of them is responsible for any errors in the result; that's either on Reichenbach, or on us.

Frederick Eberhardt and Clark Glymour

# Introduction

Frederick Eberhardt and Clark Glymour

## 1. Notes on the Translation

Reichenbach's original German dissertation is presented concurrently with the translation on the opposite page. This gave us a little more flexibility to break Reichenbach's long and complex German sentences into more manageable pieces and to change some of the punctuation to improve clarity. It also gave us a little more freedom in the translation of particular terms. In cases where we were aware of a particular translation Reichenbach himself later used for his technical terms or where standard translations exist, we tried – for better or for worse – to follow those, e. g., *Zuordnung* as *coordination* or *Anschauung* as *intuition*. In the case of *Spielraum* we decided that *event space* is the most appropriate translation, although we realize that the term has been translated by others elsewhere as *range* or *field*, and that Reichenbach uses several variations of the term on pages 7–10 of the dissertation (p. 46-52 here) in discussing von Kries's work. For other German terms, which each correspond to a variety of different English terms, we attempted to choose the most suitable translation for the particular context. So, for example, the German *Satz* may become a *theorem*, a *proposition* or a *claim*, while *Größe* may be a *variable*, *value* or *size*.

The original page-breaks are marked by vertical bars in the text in their exact original position in the German and in the closest sensible approximation in the translation. In both languages, the original page number is included in the margin. In the dissertation text we maintain page references in accordance with the original page count, but include the page number of the referent in the current volume in angle brackets. Editor's footnotes in the dissertation are distinguished from original footnotes by an index starting with a capital "E".

Errata, corrections and comments, and – when available – other material of interest will be collected on a page linked from the first author's homepage.

## 2. Historical Background on the Dissertation

Hans Reichenbach defended his doctoral dissertation at the Friedrich-Alexander University Erlangen on March 2, 1915. His advisors were Paul Hensel and Max Noether. Paul Hensel, who had written his dissertation on Kant, accepted him in Erlangen, but required a mathematician to join the committee. Max Noether, a senior algebraic number theorist and chair in Erlangen (and father of Emmy Noether), filled this position. Each took responsibility for one part of the dissertation and together they accepted the dissertation.

The dissertation was published a year later[1] in the *Zeitschrift für Philosophie und philosophische Kritik* by Johann Ambrosius Barth publishers, Leipzig. At the time of the defense, Reichenbach was 23 years old and had studied with Ernst Cassirer, Max Planck, Carl Stumpf and David Hilbert – among others – in Stuttgart, Berlin, Munich and Göttingen. In the spring of 1914 Reichenbach had visited Paul Natorp with a recommendation from Cassirer in an attempt to write his dissertation with Natorp. To Reichenbach's disappointment, Natorp declined because Reichenbach did not know enough old languages.[2] Reichenbach then wrote most of his dissertation on his own in Göttingen in May and July 1914, just before the first world war. He completed the last chapter in September of 1914 after a few weeks as a soldier. However, finding adequate advisors and the later publication of the dissertation turned out to be difficult. Reichenbach sent the dissertation to Georg E. Müller, who declined to accept Reichenbach with the comment "Do you really believe that science is furthered, when a young man comes along and develops a new scientific theory?"[3] Hensel, together with Noether, finally accepted the dissertation, but it seems as if Reichenbach was unimpressed with his advisors. In a letter[4] to Elizabeth Lingener (his first wife) he comments with annoyance on Noether and suggests that Hensel is not really interested in any of the content of his dissertation.

For publication, Reichenbach sent the manuscript to the *Kant-Studien*, where it was rejected for lack of space. The *Zeitschrift für Philosophie und philosophische Kritik* finally accepted it for publication, but Reichenbach had to pay 300 Marks for the dissertation copies – apparently his father helped out.[5]

## 3. Dissertation

The dissertation consists of 79 pages in four chapters. The first chapter lays out the main aims of the dissertation and relates it to work by Adolf Fick, Carl Stumpf

1. The standard reference for the publication is 1916. Reichenbach, however, mentions 1917 as the date of publication in his autobiographical notes; Reichenbach (1891-1953), ASP/HR 044-06-22.
2. Reichenbach (1891–1953), ASP/HR 044-06-22.
3. Reichenbach (1891–1953), ASP/HR 044-06-22.
4. Reichenbach (1891–1953), ASP/HR 041-08-09.
5. Reichenbach (1891–1953), ASP/HR 044-06-22; and Gerner (1997).

and Johannes von Kries. In the second chapter, the most technical part of the dissertation, Reichenbach identifies the key assumption for a probability distribution using a probability machine as illustration. Reichenbach shows how the same notion of probability applies to games of chance and the theory of error. The third chapter contains a "transcendental deduction" of the principle of lawful distribution, a convergence guarantee for empirical frequency distributions. Finally, the fourth chapter places the results of the dissertation in a philosophical context and discusses how probability claims about reality should be understood.

The dissertation is followed by a bibliography and a very short resumé of Reichenbach's life and education. Originally, the dissertation was supposed to include a mathematical appendix. In a handwritten draft of the dissertation available in the Reichenbach Collection at the University of Pittsburgh,[6] Reichenbach indicates that he had already completed some preliminary work for the mathematical appendix, but a note that appears to have been added later states that it was decided that the mathematical appendix should be worked out in a document separate from the dissertation for the mathematician Felix Bernstein. We have been unable to recover any draft of the mathematical appendix and are not sure which later publication might contain such a work. It seems possible that the work was never completed, since Reichenbach attended Albert Einstein's lectures on the theory of relativity shortly after the completion of his dissertation. Einstein's lectures led Reichenbach to completely revise his view on the possibility of synthetic a priori judgments.[7]

Reichenbach's handwritten dissertation draft also contains a more extensive bibliography. In addition to the references he retained in the printed version of his dissertation, his draft bibliography contains the following:

1. L. Boltzmann (1896), *Gastheorie*, Volume I & II, I. H. Barth, Leipzig
2. H. Brunn (1893), (1892), *Über ein Paradoxon der Wahrscheinlichkeitsrechnung* Sitzungsbericht der philosophisch-historischen Klasse der königlich bayerischen Akademie der Wissenschaften zu München
3. J. F. Fries (1842), *Versuch einer Kritik der Prinzipien der Wahrscheinlichkeitsrechnung*, Vieweg, Braunschweig
4. R. Laemel (1904), *Untersuchung über die Ermittlung von Wahrscheinlichkeit*, Dissertation, Zürich
5. P.-S. Laplace (1812), *Théorie analytique des probabilités*, Courcier, Paris
6. K. Marbe (1899), *Naturphilosophische Untersuchung zur Wahrscheinlichkeitslehre*, Leipzig

6. Reichenbach (1891–1953), ASP/HR 025-04-01.
7. See Section 5, p. 22 below, on Reichenbach's 1927 review of his dissertation.

7. A. Meinong (1890), *Besprechung von Kries, Wahrscheinlichkeitsrechnung*, Göttinger gelehrte Anzeigen, No. 2

8. A. Meyer (1879), *Wahrscheinlichkeitsrechnung*, Teubner, Leipzig

9. O. E. Meyer (1899), *Die kinetische Theorie der Gase*, Maruschke und Behrendt, Breslau

10. S. D. Poisson (1841), *Lehrbuch der Wahrscheinlichkeitsrechnung*, Übersetzung mit Zusätzen, Meyer, Braunschweig

11. W. Windelband (1870), *Die Lehren vom Zufall*, Berlin

12. R. Wolf (1893), *Schriften der naturforschenden Gesellschaft in Bern 1849-1851, 1853*, Naturforschende Gesellschaft in Zürich, 38 (1893), 26, 27 (1881-1883)

It is unclear why Reichenbach shortened his bibliography. The Reichenbach archive has kept handwritten notes on several of the authors above that seem to predate publication of his dissertation (e. g., Windelband, Laemel, Marbe and Meinong).[8] On Boltzmann, extensive notes are preserved. Reichenbach must have decided to delay their consideration, since he mentions in the last chapter of his dissertation that a philosophical analysis of Boltzmann's theory of gases is something future research should consider.

### 3.1 Synopsis

Reichenbach has two aims in his dissertation: The first is an analysis of the foundational assumptions required for an objective account of probability. He considers games of chance and the theory of error as two separate domains in which probability claims are used; he shows that they both require key assumptions illustrated by his so-called probability machine. In particular, these assumptions must ensure that the many (small) influences variously affecting the individual instances of a physical process combine in such a way that a general probability claim is true of the distribution of values of physical quantities – the "principle of lawful distribution". The technical analyses build on the works of Henri Poincaré, Felix Hausdorff and indirectly Carl Friedrich Gauss and aim to remove any subjective component from the truth conditions for probability claims.

The second aim of the dissertation is the integration of the foundational assumptions of probability theory into a more general philosophical – indeed Kantian – account of how scientific knowledge is possible. This requires the clarification of the meaning of probability claims and the conditions for their assertability. Reichenbach's foundational account of probability is not – as might be expected from his later writings – explicitly frequentist. There are some indications to that effect, but the meaning of probability claims is not identified

8. Reichenbach (1891–1953), ASP/HR 017-04.

with limits of relative frequencies. Instead, probability claims refer to probability functions whose existence is guaranteed by the principle of lawful distribution. Reichenbach maintains that he provides a transcendental proof of the existence of such distributions, and hence of the synthetic a priori status of the principle. That is, he claims to have proven that the existence of probability functions is a necessary condition for the attainment of knowledge and in that sense the principle of lawful distribution compliments Kant's synthetic a priori principle of causality. Only the combination of these two synthetic a priori principles make possible knowledge that can be represented as laws of nature.

In his dissertation Reichenbach is not entirely clear about the relationship between probability and causality. He takes the existence of causally independent trials to be one of the key conditions necessary for a probability distribution, which indicates that he may have thought of causation as more primitive than probability. (That causally unconnected events are probabilistically independent is assumed, but not elaborated or defended). Reichenbach explicitly states that probability is used in cases where there is no (direct/obvious) causal connection and hence presents at least the content of probability claims as in some sense disjoint from that of causal claims. However, he also describes the Kantian principle of causality as dealing with instances, while his principle of lawful distribution combines such instances "in the orthogonal direction".[9] He thereby seems to suggest that probability calculus works as a tool for aggregating causal instances. The separation of the two concepts becomes clearer in publications after his dissertation.

### 3.2 In More Detail

Reichenbach starts his dissertation by questioning the meaning of a probability claim. What does it mean when one says that the probability of a "2" coming up when rolling a die is $\frac{1}{6}$? Is it a claim about the world or is it a claim about our (lack of) knowledge? Reichenbach rejects the latter subjective view of probability and dismisses Carl Stumpf's thesis[10] that probability statements express a subjective state of knowledge – or in modern terms – a degree of belief. His main objection rests on the problem of determining which events have equal probability: What is the correct reference class that determines the probabilities of disjoint events? Should one assign a probability of $\frac{1}{4}$ to each of a parabola, circle, ellipse or hyperbola when considering the orbit of a comet, or should a circle and a parabola receive smaller probabilities since they form special cases of ellipses and hyperbolas? There does not seem to be any obvious criterion for such choices and such arbitrariness cannot form the basis of the scientific concept of probability. Stumpf's claim "Equal lack of knowledge is with respect to the resulting probability equivalent to knowledge of the equality."[11] is therefore unsatisfactory.

9. Reichenbach's dissertation, p. 73 [143].
10. Stumpf (1892a).
11. Stumpf (1892b).

Furthermore, Reichenbach claims that a subjective account of probability fails to provide a basis for rational expectations. A rational expectation makes a normative claim about how we *ought* to act given certain (probabilistic) facts about the objective world and therefore cannot vary from subject to subject. (Reichenbach does not cite the principle that probability should be a guide in life, but that is essentially his point.) A *rational* expectation should not express what the expectation – the degree of belief – actually is, but rather what it ought to be.

To Reichenbach, probability statements express something about the world and therefore any account of the foundations of probability has to be entirely objective. He cites Johannes von Kries[12] as providing one attempt in this direction. Von Kries claims that equal probability is assigned to events when there is no (logical) reason to privilege one event. This use of the "principle of insufficient reason" is, as Reichenbach correctly recognizes, not one that can support an objective account of probability. The assignment of probabilities should not be a function of what we *know* about particular classes of events. Hence, Reichenbach's first aim is the removal of this last remaining subjective element from a foundational account of probability theory.

The key problem for an objective account is the knowability of any probability claim about the world. Probability claims do not assert or deny the occurrence of any specific contingent future events and consequently it is not clear how they can be verified or falsified.

In chapter 2 Reichenbach considers a probability machine as a simple illustration of the problem and the phenomena he wants to account for. The probability machine consists of a gun shooting a needle or dart (much like a sewing machine, but with no thread) at a tape, covered with narrow black and white stripes perpendicular to the direction of motion of a tape that moves rapidly over two spatially separated rollers. Reichenbach proposes to generate a probability claim through the analysis of the distribution of the shots on the stripes of this tape. Reichenbach points out that if the tape is moving fast enough and the stripes are sufficiently narrow, then one will find that the number of shots on white stripes is approximately equal to the number of shots on black stripes if the stripes are all of equal width. This analysis of the distribution in terms of narrow equal-sized partitions of the event space is in modern terms referred to as an analysis of "strike ratios". The equal distribution of shots on black and white stripes forms the starting point for the first part of Reichenbach's dissertation: What conditions must the process satisfy in order for such an equal distribution to occur? Reichenbach identifies four such conditions:

1. The same process has to be repeated a large number of times, since no one would expect an equal distribution after just a few shots.

12. Von Kries (1886).

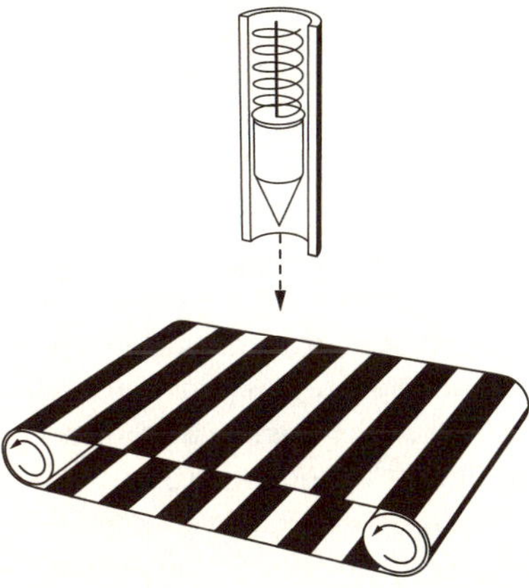

2. Initial (causal) independence of events: the shot and the moving tape must be causally independent.[13]

3. Common effect: The two causally independent events have to have some common effect, namely here the puncture of the tape.

4. Equal probability of event classes, expressed here by the equal width of black and white stripes.

Obviously, the fourth point regresses to the problem of equi-probability discussed in the context of von Kries's account. Is it possible to provide a more general assumption to ensure that the event classes have equal probability? Von Kries would say that since there is no reason to think that the shooting process privileges the white stripes to the black ones, they should be assigned equal probability. Symmetry considerations would suggest that it is the equal width of the stripes that ensures equal probability. But in the example of a weighted die, geometric symmetry would not ensure equi-probability of all the sides – there are conflicting symmetries. Reichenbach solves this problem in a different way: Scoring, say, a punch on a white stripe as 0 and and a punch on a black stripe as 1, he requires that for any continuous interval of stripes the ratio of hits (1) to misses (0) is roughly constant (not necessarily equal) and converges (as the width of the stripes decreases and the number of trials increases) to a Riemann integrable function. Reichenbach is not entirely clear on the precise nature of the convergence (see Reichenbach's dissertation, pp. 22–23 and 35 [69, 71 and 89]). In Poincaré, whom Reichenbach cites on this matter, the limit is presented in terms of a decrease in the width of the stripes, not in terms of an increase

13. Note that here Reichenbach only demands the causal independence of the tape and the shots (Reichenbach's dissertation, p. 17 [61]), but does not mention the causal independence between shots, although this assumption is evident and crucial in his later reasoning. He only writes that the shots must constitute the "same process", by which he means that they only differ in the location in space and time.

in the number $N$ of trials.[14] But there is always an implicit idealizing assumption that the number of trials (hits on the tape) is sufficiently large to allow an arbitrary reduction in the stripe width. This idealization provides a sufficient condition for the equal distribution, but it makes the task of inferring the probability distribution from the empirical frequency distribution more complicated. It requires that for any $\epsilon$, there is a number $N$ of trials such that the empirical frequency is within $\epsilon$ of the true probability. But how can this $N$ be known so that one can infer the probability distribution from the empirical frequency distribution? Reichenbach returns to this question in chapter 3 and 4.

For now, Reichenbach has specified a continuity constraint that the empirical distribution approximates. A probability of $\frac{1}{2}$ is then simply a 1:1 strike ratio, which results from the physical equality of the widths of the stripes and their (nearly) equal number on any physical interval of stripes. Part of Reichenbach's point, made again soon after by Marian Smoluchowski, is that the strike ratio does not depend on the velocity of the paper strip along the rollers, or the velocity with which the needle or dart is fired, so long as these are not in phase relative to the stripes.

Reichenbach claims that while assumptions 2 and 3 are only necessary to make the variation of the shooting process visible, the fourth point indicates that the *existence of a probability function* is a necessary condition for a stable frequency distribution. Hence, we see that two types of conditions are necessary for the applicability of probability theorems. The first type, of which condition 4 is an instance, are metaphysical principles that specify some lawfulness that distinguish natural processes for which probability claims are correct. The second type of conditions are requirements necessary to make this underlying lawfulness perceptible. The initial causal independence of events of the two types, their eventual dependence and a counting schema are such conditions and they are – according to Reichenbach – all empirically verifiable.

The conclusion Reichenbach draws from his discussion of the probability machine is that the convergence of the strike ratios from causally independent trials towards the value of a Riemann integrable function can provide an objective basis for the assignment of probabilities that dispenses with von Kries's principle of insufficient reason. The *existence* of such a Riemann integrable function is at this point left as an unsolved problem to which Reichenbach returns in chapter 4.

So far, then, the argument looks something like this: Reichenbach starts from the point of equal distribution of shots on black and white bars, i. e., a strike ratio close to 1, and attempts to find sufficient conditions for such an effect to occur. He suggests the four listed above, which divide into two types: Those that are necessary to make the variation of the process perceptible (2 and 3 – and probably 1 as well, although he does not state that explicitly) and those that are metaphysical principles (4). He provides some argument for each assumption,

14. Poincaré (1912).

but he does not give a proof for their joint sufficiency, nor does he address how much these assumptions depend on each other. The reader is largely supposed to be persuaded by the intuitive insight the probability machine allows. The account does not yet give any justification for the existence of a probability function (point 4), nor does it explain precisely what information the empirical distribution contains about the probability distribution and how the true and the empirical distribution are related – other than by some form of convergence.

Given this set-up for the discussion of probability, Reichenbach turns to games of chance as a general application of probability claims. He emphasizes the difference between an allocation of equal probabilities based on geometric considerations – such as the assignment of a probability of $\frac{1}{6}$ to each face of a die since there are six equal sized faces – and the true probability distribution over the events, which may be quite different, since the die might be weighted. Reichenbach points out the analogy between games of chance and his probability machine – best illustrated by the red and black numbers of a roulette wheel. He states explicitly that *symmetry* considerations should *not* lead us to the equi-probability of particular events. The basis for such inference should rather be the existence of a function with a continuous integral that is approximated by the physical process in question (e. g., turning the roulette wheel).

Reichenbach extends these ideas to processes involving multiple variables. The discussion of composite probability for two independent events is fairly straightforward using an argument based on multivariate strike ratios. Reichenbach again takes causal independence as a primitive. Probabilistic independence follows from causal independence since causal independence implies a distribution that is invariant under intervention,[15] i. e., an intervention on one variable does not change the marginal distribution of the other variable. Reichenbach provides no argument or more general principle for this probabilistic invariance under (causal) intervention – he is missing a notion such as the causal Markov principle that connects probabilistic and causal notions.[16] But given this invariance under intervention, it follows that the joint distribution over two causally independent events factors according to the marginals. Thus, a composite probability distribution for independent events is given when there exists a distribution function that governs the events according to the invariance just described. He does not extend this notion to joint distributions over dependent events.

Reichenbach then proceeds to argue that this notion of joint probability of independent events is informative about the approximation of the empirical frequency distribution to the true probability distribution. Since empirical distributions are based on finite evidence we can never know when the empirical distribution correctly represents the true probability distribution. Reichenbach

15. Reichenbach's dissertation, p. 33 [85].
16. A first formulation of the Markov principle can be found in *Structural analysis of multivariate data: A review*, Kiiveri and Speed (1982).

claims that we can speak of a probability distribution if for a sequence of events $\langle x \rangle = x_1, x_2, \ldots$ we have a sequence of functions

$$\begin{aligned} &\phi_1(x_1) \\ &\phi_2(x_1, x_2) \\ &\ldots \\ &\phi_r(x_1, x_2, \ldots, x_r) \end{aligned}$$

such that given any arbitrarily large $r$ and for all $0 < i \leq r$, and for any arbitrarily small $\epsilon$ there is an $N$ such that

$$\left| \frac{h^{(i)}}{N} - \int_{x_1}^{x_1+\Delta x_1} \cdots \int_{x_i}^{x_i+\Delta x_i} \phi_i(x_1, \ldots, x_i)\, dx_1 \ldots dx_i \right| < \epsilon$$

and the same $N$ holds for all functions $\phi_1$ to $\phi_r$. $h^{(i)}$ is the number of instances that actually fall within the interval.[17]

The basic idea seems to be that Reichenbach considers a sequence to form the basis of a probability distribution if each instance in the sequence is drawn from its individual (marginal) distribution and each tuple of instances is drawn from a joint probability distribution that factors into the marginals. If the empirical distribution approximates the probability distribution in this sense, then we can apply the laws of probability.[18]

The presentation is reminiscent of notions of the convergence in probability underlying the weak law of large numbers. However, what Reichenbach presents is a notion of convergence in sample size. He is missing a probabilistic measure quantifying this convergence, which would be informative for a quantitative rule for judgment of convergence and for the measure of divergence between probability distributions. His criterion would have to be adjusted to something resembling: For all $\epsilon$ and all $i$ up to some arbitrary $r$

$$\lim_{N \to \infty} P\left( \left| \frac{h^{(i)}}{N} - \int_{x_1}^{x_1+\Delta x_1} \cdots \int_{x_i}^{x_i+\Delta x_i} \phi_i(x_1, \ldots, x_i)\, dx_1 \ldots dx_i \right| < \epsilon \right) = 1$$

Reichenbach was aware of the weak law of large numbers, known then as the Bernoulli theorem.[19] However, these notions of convergence of random variables would not help solve his problem since they are based on and assume a

17. To improve clarity, the notation here is slightly modified from the original.

18. In work on the order of probability sequences by other authors, e. g., Richard von Mises (1919), sequences susceptible to treatment under the laws of probability must be random, which means (among other things) that the sequences are invariant (in terms of event probabilities) to any subsequence selection rule whose selection of any particular sequence item only depends on the event values of items prior in the sequence and the index of the item to be selected. Reichenbach's condition here involving a sequence of one- to $n$-ary functions representing the marginal distributions of one- to $n$-ary tuples seems to be doing a similar job, even though the condition is a distributional one as opposed to a restriction on subsequences.

19. See Reichenbach's dissertation, pp. 12, 68–69 [55, 135–137]. In his notes, which are thought to be from the time of his writing of his dissertation, he has a reference to Pois-

notion of probability (both as measure of convergence and in the form of independent trials) that he is trying to capture with his rule.

One may see in this discussion the roots of Reichenbach's later frequentist account of probability. The description of the type of sequences of trials that are amenable to the laws of probability is developing and the limiting notion is not quite precise yet. In particular, Reichenbach does not identify a probability claim with a limiting frequency. In fact, in later chapters he returns to this problem and states that the existence of such a probability function is a synthetic a priori. Hence, a second interpretation is perhaps more accurate: If there is a true probability distribution over a particular variable, then causally independent trials will lead to a convergence of the empirical distribution to the probability distribution. The claim here may be understood as conditional, purposefully not addressing the knowability or semantic issues of probability claims, and in the spirit (though not the letter) of the law of large numbers. This claim is not as trivial as it may seem, since it provides for an account of probability in terms of causal independence.[20] Whether such an approach is a good one is a different question, but at least it is not circular. In any case, this section is arguably the greatest source of technical problems Reichenbach addresses in his later work.

The second field of application Reichenbach considers is the theory of error. According to Reichenbach, the theory of error is based on the realization that exact measurement is impossible in principle. On the one hand exact measurement is a psychological problem due to limited perceptual resolution, but on the other hand it is a problem of nature: there are no closed systems in nature; any process is always determined by an infinity of influences. The measurement problem then amounts to the question: How does one pick the correct value among all the measurements, each of which represents an aggregation of an infinity of influences?

This section of the dissertation is based heavily on work by Poincaré and Hausdorff, who build on Gauss. Reichenbach discusses the Gaussian distribution in detail and points out that it is a special instance of a more general framework of error distributions (as shown by Hausdorff). He identifies three requirements of the Gaussian distribution:

1. The frequency of elementary errors, i. e., of the measurement error of one particular instance of an event, is given by a probability function.

son; Reichenbach (1891–1953), ASP/HR 017-04-21. The strong law of large numbers (convergence with probability 1) was known (for Bernoulli trials) to Borel in 1909, but only generally proved by Kolmogorov in 1930. It seems unlikely that Reichenbach was aware of Borel's result when writing the dissertation.

20. It may also be noted that Reichenbach retains this conditional view of probability in his later writings (though without the causal aspect).

2. Elementary error distributions are combined according to the theory of composite probability.[21]

3. Many "independent" elementary errors of the same magnitude have to co-occur.

Reichenbach points out that the third assumption distinguishes the Gaussian distribution, since it is not satisfied for error functions in general. But the first assumption embodies the assumptions already made earlier for the probability machine. Here Reichenbach attempts to pull together the various applications of probability theory in different fields (games of chance, theory of error) and show that they all face the same basic philosophical question:

*How could one know that the appropriate probability function exists?*

This is the key epistemological question of any foundational account of probability and it is the second issue Reichenbach addresses in his dissertation. The technical account he has provided so far must be integrated into the larger philosophical framework and semantic questions and questions of assertability have to be addressed.

The epistemological problem can be summarized as follows: Since we have no empirical evidence to substantiate claims that go beyond our finite experience and we certainly have no experience to support claims about an arbitrary number of events, the verifiability or falsifiability of a probability claim seems impossible. Reichenbach draws the analogy between causal and probability claims: Kant's causal principle provides epistemic support for causal claims. Its content is based in experience, but its general applicability is not empirically founded – it cannot be tested or falsified, but it is a necessary condition for causal knowledge.

Reichenbach wants to provide a foundation for probability theory similar to Kant's transcendental deduction of the principle of causality. He provides this in chapter 3 and it is – in short – as follows: Empirical laws in science make a claim regarding the validity of a particular mathematical structure in nature. They provide an approximation in the sense that the essential influences are specified in the equations bar some small errors. Physical knowledge therefore depends on the ability to specify numerical approximations in order to describe laws of nature that specify the approximate validity of certain constants for a class of events, e. g., the gravitational constant.

Causal knowledge is insufficient to make claims about the real values of physical quantities, since there is an infinity of causal influences possibly perturbing the true deterministic connection between a cause and an effect as described by an equation (or solution to an equation) of physics. Hence, in order to

21. It should be noted here, that Reichenbach's theory of composite probability does not discuss in detail how to deal with joint probabilities of *dependent* events.

obtain knowledge we need another principle in addition to that of causality that works as an aggregation procedure over several instances of causal connections, a further principle that guarantees that there is a general law about distributions of values in space and time: this role is filled by the principle of lawful distribution. So, since causal knowledge is possible, this principle is necessary, hence it is synthetic a priori. It is not an empirical principle, since it defies empirical verifiability, and it is not a logical principle, since it cannot be derived from the principles of logic.

Reichenbach claims to have provided a transcendental deduction, analogous to Kant's, for the existence of a probability function, i. e., a convergence limit for the empirical distribution. Without the existence of probability functions, no knowledge as represented by the laws of nature would be possible.

This principle ensures that the claims of the probability calculus can be made with certainty: It is, according to Reichenbach, what is certain in order for something else to be probable. The outcome of an individual case is probable, but convergence of the distribution is certain. If convergence does not occur, then the basic assumptions for the probability distribution were not satisfied (e. g., causal independence of trials). Lack of convergence, however, does not disprove the principle.

This last point is, of course, the pitfall (or perhaps, the pratfall) for Reichenbach's account. It begs the question, since it leaves open the whole issue of how one can be sure that the observed convergence in the empirical distribution is representative of the true probabilities. Reichenbach is acutely aware of this problem and admits that the principle cannot be disproved, but points out that this applies similarly for the principle of causality. Furthermore, he points out that the impossibility of falsification does not make the principle of lawful distribution vacuous, since it contains positive content: A sequence may converge and later diverge again, but the claim that is made with certainty is that for any $\epsilon$ there will be an $N$ for which convergence within $\epsilon$ is final. It is hence not equivalent to the claim that any distribution may be achieved. And this aspect provides – according to Reichenbach – a basis for rational expectation. It seems a rather weak assurance.

Reichenbach concludes his dissertation by placing his thesis in the larger philosophical context: Adolf Fick had claimed that the calculus of probability is synthetic a priori. Reichenbach's dissertation proves that the principle of lawful distribution is an objective law of nature, which implies that the theorems of probability calculus hold necessarily for reality. Together, these studies show the parallel between geometric and probabilistic laws. They provide an explanation why analyses in terms of probability provide the basis for knowledge.

He states as his further research the aim to give a philosophical account of Maxwell-Boltzmann statistics in thermodynamics.

## 4. Influences on the Dissertation and Reichenbach's References

References to earlier work divide according to the two main parts of the dissertation. On the technical side (chapter 2) his work is most influenced by Poincaré and to a lesser degree by Hausdorff, who both draw on Gauss, while his philosophical views (chapters 1, 3 and 4) are Kantian in a general sense, but derive insights on judgments of probability from the work of Fick, Stumpf, and von Kries.

In the following we will focus the discussion on Fick, Stumpf, and von Kries and a brief summary of the technical references, as Kant's work is discussed more competently and in more detail elsewhere.

### 4.1 Adolf Fick

Reichenbach takes the central thesis question – the relation between probability theory and reality – from Fick.[22]

Fick takes probability to refer to hypothetical judgments. In such a hypothetical judgment the antecedent describes imprecisely or incompletely a set of causal conditions which may lead to the event specified in the consequent, e. g., When I throw a coin, then it comes up heads. A numerical probability refers to such a hypothetical judgment in the sense that it specifies the fraction of possible causal constellations in the antecedent that necessitate the event specified in the consequent.[23] Accordingly, probability is closely related to causality for Fick, but his account does not undermine the view that causal relations are necessary.

It remains entirely opaque how one is supposed to establish which of the causal constellations covered by the antecedent necessitate the outcome or how one should count these constellations in order to determine the numerical probability value. Fick provides no advice on how to determine equi-possible events, which surely any such account of probability needs to describe.

Apart from this foundational problem, Fick is aware of the more general problem of the verification and falsification of judgments of probability. He takes the fact that empirical evidence is never sufficient or conclusive for such verification or falsification of judgments of probability as an indication that these judgments must have a synthetic a priori nature.[24] If claims of probability are synthetic a priori, then the question arises why reality appears to approximate these principles of probability. Fick considers this to be an open metaphysical question.[25] It is this question that Reichenbach cites and uses as the core question of his thesis.[26] Fick is the only one of the three philosophers and probability

22. Fick (1883).
23. „Die Wahrscheinlichkeit eines unvollständig ausgedrückten hypothetischen Urtheils ist der als echter Bruch dargestellte Theil des ganzen Bereichs der Bedingungen, an dessen Verwirklichung der im Nachsatz ausgedrückte Erfolg *notwendig* geknüpft ist.“ Fick (1883), p. 12, emphasis in original.
24. Fick (1883), p. 46.
25. Fick (1883), p. 46.
26. Reichenbach quotes Fick's comment on p. 13 [57] of his dissertation.

theorists who explicitly refers to Kant for an epistemological foundation of probability theory. Even though Reichenbach relies relatively little on Fick's work, it must be assumed that he derived substantial motivation for his more thorough Kantian analysis from Fick's ideas.

Instead of focussing on Fick's work, Reichenbach places his dissertation in the context of the work of von Kries and Stumpf – both of whom refer to Fick. Reichenbach takes them to be instances of two diametrically opposed views on the interpretation of judgments of probability, neither of which he wants to fully endorse in his dissertation.

According to Reichenbach, von Kries exemplifies the closest view available to an *objective* interpretation of probability, the only sticking point being the principle of insufficient reason. In contrast, Stumpf's account is taken as an example of a subjective interpretation of probability.

## 4.2 Johannes von Kries

In *Die Prinzipien der Wahrscheinlichkeitsrechnung* (1886) von Kries endorses Fick's approach as the right start to an objective understanding of probability,[27] but points out that without the principle of insufficient reason, Fick's account cannot explain how objective probabilities can be assigned uniquely.[28] His work can therefore be seen as a development of Fick's view.

Von Kries takes judgments of probability to describe some objective feature of the world and to support what he refers to as a "logical" view of rational expectations (as opposed to a practical or subjective one). Judgments of probability are, on his view, based on and derived from a set of original events – we will refer to them as *ur-events* here – that *are* or *are deemed to be* (this is where the problem lies) equi-possible. Given this foundation, the numerical probability theory rests on two principles: the principle of event spaces, which says that events that contain equal and indifferent ur-events are equally probable; and the principle of lawfulness, which essentially amounts to a principle of uniformity of nature.[29]

The problem is – as von Kries is well aware – how to determine the set of equi-possible ur-events without falling into a subjective view of probability. On the one hand he wants probabilities to be objectively valid, but on the other hand he concludes that the equi-possibility of events can only be determined by the lack of a sufficient reason to believe the contrary. That is, equi-possibility is determined by the principle of insufficient reason: *Two events are equi-possible if at the current state of knowledge there is no reason to consider one of the events more likely than the other.* In order to support a "logical" or objective view of probability this determination of equi-possibility must be free of any arbitrariness, i. e., there must be a unique partition of ur-events that are equi-possible and that form the basis of composite probabilities.

27. Von Kries (1886), p. 284.
28. Von Kries (1886), p. 285.
29. See von Kries (1886), chapter 7.

In modern terms, von Kries's views resemble most closely those of an "objective Bayesian" and many of his aims and difficulties become readily understandable when thought of in this context.

The tension between the objectivity of the probability judgment and the seeming subjectivity of its epistemological foundation is present throughout von Kries's account and it is the reason why Reichenbach takes issue with the principle of insufficient reason in his dissertation.

According to von Kries, any interpretation of probability judgments has to provide an account of how to determine equi-possible events. Von Kries considers and dismisses several different views. We will only sketch his arguments here, but the reader should take note of how many of the same arguments reoccur in Reichenbach's dissertation:

1. A subjective choice of equi-possible events is inadequate since that would leave probability judgments open to arbitrariness, which they are not – at least not if one wants to talk about probabilities in the sciences. On the subjective view, there is no reason to believe there is a unique partition of the event space into equi-possible ur-events, and further, it would leave judgments of probability subject to factors generally considered alien to them, e. g., desires, utilities, expectations and psychological limitations.

2. Equi-possible events cannot be determined by reference to physical features, e. g., equal geometric (or other) partitions of the relevant event space, since (i) the reference metric used for the division may not be unique, again implying arbitrariness, and (ii) equal partitions on whatever metric is chosen may nevertheless not correspond to equi-possible events.

3. We cannot take probabilities of events to constitute the observed frequencies of the events in our experience so far, thereby letting probabilities resemble our current state of knowledge. We consider a probability of 1 to represent certainty, which it clearly is not on this account. A probability of 1 would only represent the claim that so far we have only seen one type of outcome, but that is not equivalent to being certain that there only is one type of event.

4. Perhaps most notably, von Kries considers probabilities to represent limiting frequencies of events in a sequence of trials (though he does not use this exact terminology). Von Kries dismisses such an interpretation, since (i) we have no experience of infinite sequences, (ii) we generally do not explicitly assume the existence of a limit of the relative frequencies in a sequence of trials, and (iii) such an account would clearly have problems with the verification and falsification of any probability judgment.[30]

30. This last consideration is noteworthy for two reasons: First, it lays out many of the problems Reichenbach struggles with in his frequentist account much later in life. And second, even given this criticism, von Kries does later in the text effectively return to some-

Given the unsatisfactory nature of these candidate interpretations, the key question for von Kries is how we can determine conditions for knowledge that support the principle of insufficient reason. Ultimately, his argument is unsatisfactory, in particular with regard to the principle providing some type of non-subjective support for judgments of probability. It hinges on the notion of a "*free formation of expectation*",[31] which is supposed to enable some objective, independent assessment of equi-possibility. But von Kries's analysis of this notion is vague, as he admits himself: "If we start from the assumption that the probabilities associated with games of chance are the correct ones, then we would have reason to demand a general and strict proof that these probabilities really express the ratios of indifferent original event spaces. But it goes without saying that in this case, as a real situation, a mathematical or purely logical proof can never be given."[32]

It is nevertheless worthwhile to dwell on von Kries's analysis a little longer, since he takes many of the approaches Reichenbach develops and uses himself in his dissertation, but reaches slightly different conclusions.

Von Kries aims to find equi-possible events by backtracking in the causal history of events until we hit an original event space for which each individual event is equally likely or until any further backtracking does not seem useful. This space of equi-possible "ur-events" then defines the probability of any resulting composite events. The problem is how one is to go about discerning these original event spaces.

In the case of games of chance, von Kries argues that an analysis in terms of strike ratios[33] provides the type of indifferent original ur-events (namely hitting a black or a white stripe) to which the principle of insufficient reason can be applied. He does not explicitly have Reichenbach's notion of convergence to a Riemann integrable function but the general idea is there.[34] He uses a shooting device for illustration which Reichenbach essentially copies with his probability machine. Von Kries argues that the continuity of the distribution function is insufficient to obtain the equal distribution of hits on black and white stripes.[35] The principle of insufficient reason is still required to ensure that we have no reason to believe that any part of the shooting device or the alignment of the stripes results in a preference of hits on one color, i. e., we have to be convinced that the shooting process is independent of the stripe color, and that can only be achieved by the principle of insufficient reason.

If this reconstruction of von Kries's argument is correct, then Reichenbach's critique of this account[36] presents a crucial insight about strike ratios,

thing rather closely resembling a frequentist interpretation, with the principle of insufficient reason filling the gaps.

31. See Reichenbach's comments of this notion in his dissertation, p. 8 [49].
32. Von Kries (1886), p. 60.
33. Von Kries (1886), p. 50.
34. Von Kries (1886), p. 64.
35. Von Kries (1886), p. 74.
36. Reichenbach's dissertation, p. 30 [81].

but does not replace the principle of insufficient reason as von Kries uses it: Von Kries demands that ur-events have equal probability and that we can find such ur-events by considering strike ratios. Reichenbach's point is that it is the continuity of the probability distribution being approximated by the empirical frequency distribution rather than the equality (in the limit) of the number of hits on black and white stripes that matters for the foundation of probability. That is, if the alignment of stripes were such that white stripes are three times the width of black ones, then we would still be able to speak of a probability distribution over the hits on the stripes, namely 3/4 white and 1/4 black, but it would not be an equal distribution.[37] So, if we take strike ratios to be ur-events, as von Kries does, then there is no necessity for an equal (constant) distribution over the ur-events. But it is a mistake to conclude from here, as Reichenbach does, that his principle of lawful distribution is superior to von Kries's principles.[38] Von Kries's principle of insufficient reason also embodies the assumption about the independence of trials and the (prior) independence of the shooting process from the stripe color, which Reichenbach just stipulates. The difference in the two views might best be seen in terms of a technical vs. a pragmatic account: Reichenbach is aiming to specify an idealized account of probability as one may find it in the sciences. In contrast, von Kries is concerned with the pragmatics of establishing a probability claim, and realizes that in order to make the idealizing assumption, one realistically has to resort to the principle of insufficient reason.

This is where the "free formation of expectation" comes in. Essentially, von Kries argues that it can be formed when considering a large number of independent identically distributed trials. He does not use these exact terms but refers to sequences of trials of similar type,[39] and with constant conditions. Trials must be independent and the sequence must be extendable indefinitely.[40] It is not entirely clear what he means by trials of similar type or by constant conditions, but the gist seems to be that the trial takes place in roughly the same causal context and that over a sequence of trials there is no reason to believe that conditions affecting the outcome are significantly different in each trial.

His discussion of the independence of trials is more detailed: Von Kries considers sequences of *causally* independent trials as support for equi-possible events,[41] but dismisses this approach (unlike Reichenbach later) as inadequate. Like Reichenbach, he believes it is impossible to have two sequences of events that are completely causally independent.[42] Instead he argues that we treat trials *as if* they were causally independent[43] when we have no reason to believe otherwise. Basically, he is struggling to specify in detail some further notion that

37. The second paragraph on p. 20 [67] of Reichenbach's dissertation makes this point particularly clear.
38. Reichenbach's dissertation, p. 31 [83].
39. "gleichartig", von Kries (1886), p. 103.
40. Von Kries (1886), p. 104.
41. Von Kries (1886), p. 38.
42. Von Kries (1886), pp. 38–39.
43. Von Kries (1886), p. 83.

would amount to a definition of randomness, but it remains – unsurprisingly, given the later difficulty von Mises and others had with this notion[44] – somewhat vague, just re-expressing the pre-theoretic notion: a trial has a random outcome (though this term is not used) when there is "very significant uncertainty"[45] about the outcome and when the outcome can be generated in many different ways, i. e., there are many factors that contribute to the outcome.[46]

This latter point is again an issue we find in Reichenbach's dissertation in the context of the theory of error: Judgments of probability are relevant in cases where there is a large number of small factors that contribute to a outcome.

The core difference in Reichenbach's account is that von Kries argues that in the end the determination of equi-possible events is a matter of argument by analogy and inductive inference.[47] Since there cannot be any logical or mathematical proof of the equi-possibility of events, the best we can hope for is a confidence in our judgment that eliminates any reasonable doubt. It remains unclear what exactly constitutes such a confidence and how one could test for it or notice when it is false. Von Kries sees the appeal of an interpretation of probability in terms of limits of relative frequencies, but is very sensitive to the problem of verification and falsification of any probability claim based on such an account[48] and what the appropriate empirical support would be.

Reichenbach does not take this line of argument, since he wants to remove all subjective elements from the interpretation of probability. He instead tries to provide a transcendental deduction that probability judgments can be based on a synthetic a priori principle about the existence of limits of relative frequencies in sequences of causally independent trials.

So, while Reichenbach shares most of von Kries's analysis of the foundation of probability, he does not share his views on the epistemological basis for an objective interpretation of probability. He wants to turn von Kries's "objective Bayesian" account of probability into a truly objective account.

### 4.3 Carl Stumpf

In contrast to von Kries, Carl Stumpf lays out a thoroughly subjective interpretation of judgments of probability in *Über den Begriff der mathematischen Wahrscheinlichkeit* (1892a). He also takes his account to provide a treatment of probability as it is used in mathematics and refers to Laplace as the source of this view,[49] which Reichenbach, however, considers to be a misinterpretation of Laplace.[50] Stumpf's aim is to make Laplace's notion of equi-possibility more precise.

44. See subsection 6.2, p. 26 below, on the definition of probability.
45. Von Kries (1886), p. 58.
46. Von Kries (1886), pp. 68–69.
47. Von Kries (1886), p. 79.
48. Von Kries (1886), chapter 6.
49. Stumpf (1892a) p. 40.
50. Reichenbach's dissertation, p. 6 [45].

Stumpf takes issue with Fick's view by arguing that if we were able to express the antecedent causal conditions sufficiently accurately, then there would be no place for probability claims anymore. We could analyze the causal nexus and then determine exactly which outcome necessarily follows. Stumpf argues that probability has nothing to do with such a causal story. Furthermore, he takes it to be virtually impossible to ever be able to specify and enumerate all the conditions for an outcome in such a way that we would be able to determine which fraction of them necessitates the outcome. He says, "...we say that it [a coin coming up heads] has probability $\frac{1}{2}$ not because we *know* that this outcome will occur for half the [antecedent causal] conditions, and fails to do so under the other half, but rather because we *do not know* which of the two types of conditions apply in the given case."[51] A probability claim is not about the relation between antecedent and consequent of a hypothetical judgment, it is rather a claim about the lack of knowledge of the occurrence of each particular event.

On Stumpf's view judgments of probability express a state of uncertain knowledge. "We say that any event is $\frac{n}{N}$ probable if we can conceive of it as one of $n$ elements (favorable instances) among a total of $N$ elements (possible instances), of which we know that exactly one is true, but we do not know which."[52] If we have no reason to consider one outcome to be more likely than another, we assign equal probabilities to each possible outcome. Stumpf thereby avoids the whole discussion surrounding the principle of insufficient reason. He argues that a proper representation of all our knowledge allows for a use of probability that avoids the paradoxes resulting from an arbitrary choice of probability assignments.

Stumpf explains how Bernoulli's theorem can be captured by a subjective interpretation and presents in *Über die Anwendung des mathematischen Wahrscheinlichkeitsbegriffs auf Teile eines Continuums* (1892b) a response to von Kries's charge that a subjective interpretation of probability leads to conflicting probability assignments.

Stumpf's central claim, which Reichenbach later cites as absurd in his dissertation[53] is that "With respect to the resulting probability equal lack of knowledge is equivalent to knowledge of the equality."[54]

Stumpf holds that his notion of probability does not involve any assumptions or convictions about an objective world and he explicitly rejects the idea

51. Stumpf (1892a), p. 61.

52. Stumpf (1892a), p. 48, but see footnote in translation of Reichenbach's dissertation, p. 3 [43].

53. Reichenbach's dissertation, p. 5 [45].

54. Stumpf (1892b), p. 687. Immediately following this statement, Stumpf includes a caveat not included in Reichenbach's citation: "(except in the case where we are uncertain about only a single event, in which case this claim would imply that the probability becomes certainty)", parentheses in original. This caveat certainly avoids von Kries's charge that if probabilities represented experienced relative frequencies, then the experience of a single event would result in probability 1, i. e., certainty. But it remains unclear how Stumpf would resolve this case.

that probability is dependent on a notion of causality.[55] In fact, he argues that probability is particularly applicable in the absence of any causal notions and that many uses of probability involve no concept of cause and effect. To Stumpf it is the lack of knowledge about the occurrence of events, not causal aspects, that lead to an assessment of probability. Consequently, he does not consider the realization that a die is biased as involving a *correction* of the earlier probability assessment, but rather just a *change* of probability.[56]

Reichenbach does not really add anything new to von Kries's rejection of (Stumpf's) subjective interpretation of probability. It is not entirely clear just how far apart von Kries and Stumpf are in their views. Stumpf argues that if we take all our knowledge into account then we avoid paradoxes resulting from an arbitrary assignment of probabilities. Similarly, von Kries agrees that the principle of insufficient reason ultimately is based on the epistemic state of the individual (what more could a free formation of expectation be?). It therefore seems that Stumpf just explicitly bites the bullet and says that, in the end, equipossibility is subjectively motivated. The dichotomy that Reichenbach wants to draw is never illustrated with an instance where one could safely say that von Kries and Stumpf assign different probabilities to events, even though there appears to be little doubt that von Kries and Stumpf took their views to be opposed. The difference seems to be, so to speak, only philosophical.

### 4.4 Technical Sources: Henri Poincaré and Felix Hausdorff

Chapter 2, the technical part of Reichenbach's dissertation, is based mostly on Poincaré's *Calcul des Probabilités* (1912). Poincaré provides a discussion of probability in terms of strike ratios[57] and proves that the probability of hitting a white stripe is $\frac{1}{2}$ if the distribution function is continuous and differentiable and the stripe widths are "very small".[58] He shows that this holds true no matter what the form of the distribution function is (as long as it bounds a finite area with the $x$-axis). This is often referred to as the method of arbitrary functions. Poincaré also points out explicitly in *Wissenschaft und Methode*[59] that an application of probability theory to actual events requires the assumption of the existence of a probability distribution, which the real events approximate, i. e., the existence of a limit. The only justification Poincaré offers for this assumption is a rather simplistic uniformity of nature claim.[60]

There is also no discussion of the independence and identical distribution of trials. This is assumed implicitly in the presentation of the set-up (e. g., a roulette wheel), although von Kries had pointed out the importance of this assump-

55. Stumpf (1892a), p. 49.
56. Stumpf (1892b), p. 103.
57. Poincaré (1912), pp. 147–152; *Wissenschaft und Methode*, Poincaré (1914), p. 71.
58. In his autobiographical notes from 1927 Reichenbach notes that he came up with this continuity condition independently, but later found it in Poincaré's work and hence decided to cite him instead. Reichenbach (1891–1953), ASP/HR 044-06-21.
59. Poincaré (1914), chapter 4, VIII, p. 73.
60. Poincaré (1914), p. 74.

tion much earlier. In particular, the need for independence of the stripe color from the outcome selecting device is not mentioned anywhere.[61]

Both Poincaré and Hausdorff are cited in Reichenbach as references on the theory of error. They provide detailed technical discussions of the Gaussian theory of error. Hausdorff, in particular, provides a derivation of the Gaussian distribution and makes the assumptions for the derivation, which Reichenbach emphasizes,[62] explicit.[63] Hausdorff generalizes the theory of error and shows that the Gaussian distribution is a special case of a more general set of error functions (that do not satisfy all the constraints he lays out for the Gaussian case).

### 4.5 Summary of Influences

One is tempted to conclude that Reichenbach's dissertation presents nothing new: Strike ratios were discussed by von Kries and Poincaré in philosophical and technical terms, respectively. The probability machine essentially resembles von Kries's shooting device and the results on the theory of error are taken from Poincaré and Hausdorff. But while many of the ideas and results were already around, no one had brought the technical and philosophical aspects so close together (remember that Reichenbach had difficulty finding a suitably qualified advisor for the dissertation) and certainly no one had taken Fick's request seriously and attempted to resolve the epistemological difficulties of probability theory in a Kantian framework.

Curiously, if it were not for his dissertation, one could see Reichenbach's later work on the frequentist interpretation of probability judgments in *The Theory of Probability* (1935; 1949) as a direct continuation of von Kries's work. The dissertation, however, places a little Kantian warp in this trajectory.

## 5. Reichenbach's 1927 Comments on his Dissertation

Unpublished autobiographical notes are preserved from August 6, 1927 in which Reichenbach gives a brief review of his dissertation.[64] The notes show clearly the realization of the problem with the Kantian synthetic a priori and a development towards a (hierarchical) notion of convergence in probability. They also show the shift towards considering probability as primitive and the importance Reichenbach sees in his contribution to the relation of probability and causality. We include here a translation of the notes:[65]

"Results of my work from 1914:

1. I showed that the condition of equal probability can be reduced to a continuity condition – at least for a class of problems.

61. See also Michael Strevens's comment in *Bigger than Chaos*, (2003), p. 119.
62. Reichenbach's dissertation, p. 39 [95].
63. Hausdorff (1901), p. 169.
64. Reichenbach (1891–1953), ASP/HR 044-06-21.
65. Reichenbach's notes were handwritten, so there may be errors in the transcription.

2. I showed that this continuity condition does not only apply to problems in probability, but is assumed for all physical claims; without it causal claims would be vacuous.
3. I attempted to base the probability claim on a claim of certainty. (There is with certainty an $N$ such that $\epsilon < \Delta\eta$, where $\Delta\eta$ is given, whereas it is not certain whether there is an $N$ such that $\epsilon > \eta$.)[66]
4. I attempted to prove that the probability condition[67] is a synthetic a priori judgment that is necessary for all knowledge.

I now have the following to say about the above: 3 and 4 are failed attempts, 1 and 2 succeeded.

On 3, I note that the claim about the probability function could only have the following content: the events will *probably* behave in accordance with the probability function. The claim of certainty is in fact vacuous, since $N$ cannot be computed, unless the claim is a probable claim. Probability has to be introduced as foundation [*Grundlage*]. – Grelling already made this point to me in 1914, and Paul Hertz did later.

On 4, I note that only a single publication convinced me of the impossibility of synthetic a priori judgments: my own (1920).[68] There I already noted that I have to correct my work on probability. I also remarked in a symposium that the [assumption of the existence of a probability function] is *not* a synthetic a priori judgment.

On 1: Here Poincaré deserves the most merit. I discovered the condition independently, but I later found it in Poincaré, and consequently cited him as source. – By the way, I certainly do not consider it hopeless to extend the condition to problems of [?].[69]

On 2: This fact I discovered myself, I still consider it to be the most important discovery that has been made with regards to the problem of probability since Hume."

## 6. Reichenbach's Theory of Probability after the Dissertation

Following his dissertation in 1915, Reichenbach attended Einstein's lectures on the theory of relativity in Berlin. This influence became evident immediately in

66. [Reichenbach's notes do not give more detail here, but it seems that $\epsilon$ represents the difference between the empirical frequency and the true distribution, while $\Delta\eta$ is some given measure of accuracy of approximation and $N$ is the number of trials. (Footnote added by the editors.)]
67. [This refers to the principle of lawful distribution, the condition that guarantees that the frequency distribution converges to a limit. (Footnote added by the editors.)]
68. [Reichenbach is referring to *Relativitätstheorie und Erkenntnis a priori*, (1920c; 1965). (Footnote added by the editors.)]
69. [We were unable to decipher the word in Reichenbach's hand-writing, but it may be "Mischungsproblem", i. e., problems of mixtures. (Footnote added by the editors.)]

his publications and spelt the end to his Kantian views.[70] The bearing of the theory of relativity and results from quantum theory on philosophy of science and on philosophical questions in general developed into one of his main areas of interest. Nevertheless, he continued publishing on issues of probability and causality. The former culminated in his book *The Theory of Probability* published in 1935 in German. The latter is only fully addressed in its own right in a manuscript that was published by his wife, Maria Reichenbach, in 1956 as *The Direction of Time*. However, between 1915 and 1935 Reichenbach published several short papers pertaining to both topics, "Die physikalischen Voraussetzungen der Wahrscheinlichkeitsrechnung" (1920b), "Die Kausalstruktur der Welt und der Unterschied von Vergangenheit und Zukunft" (1925), "Stetige Wahrscheinlichkeitsfolgen" (1929), "Kausalität und Wahrscheinlichkeit" (1930) and "Axiomatik der Wahrscheinlichkeitsrechnung" (1932a).

In these papers, three developments of Reichenbach's probability theory can be distinguished. First, a struggle with the choice of primitives. The primacy of the principle of causality – largely due to Kant – is questioned and finally removed. Reichenbach appears to struggle with the epistemological implications of the results from quantum physics, which seem to undermine his basic view of a deterministic world where probability theory is only an aid in the face of limited knowledge; but he relegates the concerns to footnotes and does not address them yet.

Second, the definition of probability changes from one based on causally independent trials to one based on the limiting relative frequency in probabilistically independent trials. Reichenbach attempts to avoid the apparent circularity of such a definition by several attempts to define a notion of "normal sequences". This account was supposed to present an improvement on Richard von Mises's requirement of random sequences to support claims of probability, which was subject to several theoretical problems.

Third, Reichenbach develops his solution to the problem of the ascertainability of probability claims. Since the a priori nature of the principle of lawful distribution is weakened and finally abandoned, Reichenbach needs a new account ensuring convergence of the empirical distribution. He proposes a pragmatic account of how, given finite evidence, we can make probability claims that, given their definition, are about infinite sequences. He argues that the empirical distribution provides the best bet for the true distribution. The justification is purely pragmatic: If the distribution converges, then betting on the empirical distribution will get us to the truth. Reichenbach places this principle outside the domain of the probability calculus and later integrates it into an inductive logic. The principle is now known as the *straight rule*. In all of these changes, the move away from Kant is apparent. There seems to be a deep skepticism towards the epistemic status of the transcendental deduction, the foundations, and the value of the assurances it provides.

70. Reichenbach (1920c; 1965).

### 6.1 De-Kanting Primitives: Probability or Causality

Results from quantum physics challenge the assumption of a deterministic universe. Probability is then not only seen as filling the gaps where we do not have sufficiently precise measurements to record all the causes of a particular event, but rather becomes a feature of nature itself. Even if we had infinitely precise knowledge we would not discover the world to be governed by deterministic causation, instead probability relations are intrinsic features of the world. Reichenbach noted these implications (albeit in footnotes)[71] for the status and epistemological role of probability very early. It undermined his Kantian world view that singular event causation formed the basis of the universe. In addition, the whole idea of causation as a primitive came into question, since Reichenbach realized that we generally determine causal relations by some – not fully explicit – probabilistic criteria.

There appears to be a shift in Reichenbach from a Kantian view of causality of singular events to an analysis of causality as a relation between event types: Causal claims are not anymore about particular instances of events, such as "billiard ball A causes billiard ball B to move", but rather about variables that take different values in a series of instances, e. g., "air pressure causes barometer needles to move". Rather than being a micro-level feature, causality is understood to be a macro-level feature of events. This switch is explicit in his papers: In "Stetige Wahrscheinlichkeitsfolgen" (1929) Reichenbach still refers to (deterministic) causal relations obtaining at the infinitesimal micro level, whereas any more general description of higher level causation (and any human knowledge about causation) can only be probabilistic. This changes in "Kausalität und Wahrscheinlichkeit" (1930) to an account in which causation describes high-level regularities that result from an infinity of events that are only describable in terms of probabilities. Here Reichenbach identifies Boltzmann as one of the first people voicing this view.

It is apparent that Reichenbach has a chicken and egg problem during these years: What comes first – causality or probability? Which is the more fundamental concept? His dissertation provides an account of probability in terms of causality. Causation is the primitive, but probability is – just like causality – a necessary condition for physical knowledge. This view regarding physical knowledge is maintained in "Die Kausalstruktur der Welt und der Unterschied von Vergangenheit und Zukunft" (1925), but the hierarchy is reversed: Probability is the primitive and the only synthetic a priori,[72] while causality is described in terms of it. Reichenbach does not provide details of this relation. In particular, he does not give an explicit definition of causality in terms of probability. Nevertheless, this switch is bad news for his definition of probability, since it relied on causally independent trials (see next section). By the time he writes "Kausalität und Wahrscheinlichkeit" (1930), Reichenbach has reverted to claiming

71. E. g., Reichenbach (1925), p. 138, Reichenbach (1929), p. 305, and Reichenbach (1932a), p. 571.
72. Reichenbach (1925), p. 145.

that both causality and probability are fundamental necessary conditions for any physical knowledge. Finally, when he writes *The Theory of Probability* (1935; 1949) he describes the problem of providing an account of causality as unsolved. In *The Direction of Time* (1956) he starts with the claim that he will provide a reduction of causality to probability, but – since he did not complete the book – it remains unknown whether he ever had an adequate solution. All that is provided there are the time ordering and screening off conditions (principle of common cause), which Suppes later developed further.[73]

## 6.2 Definition of Probability

The concern raised by the loss of the Kantian influence is the foundation for Reichenbach's definition of probability. Without causality as a primitive, a definition of probability cannot be founded on causally independent trials. In "Die Kausalstruktur der Welt und der Unterschied von Vergangenheit und Zukunft" (1925), he simply resorts to making probability a primitive. But this does not last. The development of an account of probability in terms of limits of relative frequencies of particular sequences of events is gradual, but becomes explicit in "Axiomatik der Wahrscheinlichkeitsrechnung" (1932a). The main problem with such a definition is the risk of circularity: If one gives up the requirement of causally independent trials, then one has to argue that probability is the limiting frequency of an infinite sequence of independent identically distributed trials. But then probability is defined in terms of independence, which is – presumably – defined in terms of probability.

Von Mises (1919) suggested that a sequence of independent and identically distributed trials could best be characterized as a *random* sequence of events, that is, a sequence of events on which one cannot make any money when placing bets. Reichenbach avoids von Mises's account, since there are several theoretical difficulties in attempting to give a formal definition of this pre-theoretic notion of randomness. Two features are commonly taken to be hallmarks of randomness: First, the lack of after-effects, i. e., that the probability of a particular outcome is the same no matter what the previous outcomes were. And second, that the probability of an event is invariant under any subsequence selection rule that is dependent only on events prior in the sequence and on the index of the event to be selected. It was doubtful whether any formal accounts of this pre-theoretic notion of randomness could be given[74] and Reichenbach had pointed out how some candidates fail.[75] Reichenbach, instead, avoided randomness and formalized the notion of a "normal" sequence.[76] The definition satisfied the lack of after-effect and an invariance under a particular type of subsequence selection (regular division), but generated a class of sequences more general than the one von Mises was trying to capture. Reichenbach claims to have developed

73. Suppes (1970).
74. See Church (1940).
75. Reichenbach (1932a).
76. Reichenbach (1932a).

the notion of a normal sequence independently, but acknowledges that Arthur Copeland had suggested an almost equivalent notion of sequences of "admissible" numbers earlier.[77] In fact, Émile Borel had proved results for normal sequences as early as 1909.[78]

Normal sequences are attractive as a basis for the definition of probability, since they provide a foundation purely in terms of subsequence selection rules and limiting frequencies. However, their conditions are insufficient to characterize the lack of lawfulness von Mises wanted for random sequences: Normal sequences can be generated by a recursive rule and they may not satisfy the invariance of event probabilities under place selection procedures that are not regular divisions. Reichenbach was keen to keep the classes of sequences amenable to the laws of probability as broad as possible and hence was less concerned with these issues. In fact, in his later work, Reichenbach argues that probability applies to any sequence whose relative frequencies are convergent. This provides a very simple definition of probability, free from all the complications of subsequence selection rules and instead with convergence as the sole criterion. But it now places all the hard work on the problem of ascertainability of probability claims. If the sequence is not patterned, how and why can one establish the probability of a particular event given that one has not seen the entire infinite sequence and does not know whether it converges?

After 1935, the debate on an account of randomness proceeded largely independently of Reichenbach. Von Mises was unsatisfied with the account in terms of normal or admissible sequences. Abraham Wald proved a positive result for tests of random sequences,[79] which at least ensured that the notion of randomness was not self-contradictory. Jean-André Ville proved a negative result that there are sequences that satisfy von Mises's definition of randomness, but that converge to the limit from one side.[80] This went against the pre-theoretic idea that one cannot make money when placing bets on the sequence of events. Finally, Alonzo Church gave an account of random sequences that was more restrictive than normal/admissible sequences, but it did not quite provide the restriction von Mises was looking for, since it did not cover *any* subsequence selection rule, but rather restricted itself to effectively calculable ones.[81] On Church's account, the existence of a random sequence can be given only by a non-constructive proof, which – as Church points out – may be objectionable for a foundation of probability. It is nevertheless generally agreed that Church's formal account is as close as one is going to get to von Mises's pre-theoretic notion of randomness. Anything closer runs the risk of self-contradiction, anything weaker is susceptible to money-making bets. For more detail, Church's work and a short paper by Per Martin-Löf[82] provide a good overview.

77. Copeland (1928).
78. Borel (1909).
79. Wald (1938).
80. Ville (1936).
81. Church (1940).
82. Martin-Löf (1970).

In the English edition of *The Theory of Probability* (1949) Reichenbach refers to these developments, but largely sticks to his definition of normal sequences.

### 6.3 Solution to the Problem of Assertability

The third worry left from Reichenbach's dissertation is the question of what justification we have to take the empirical distribution as indicative of the true distribution. How can we be sure that the sequence of events is convergent (and if so, to what?) in order for probability claims to apply? The dissertation provided the principle of lawful distribution as a synthetic a priori assurance that guaranteed convergence: If we did not assume that the empirical distribution converges, then scientific knowledge would be impossible. This view gradually changes into the pragmatic account given by the so-called straight rule in *The Theory of Probability*[83]: Treat the empirical distribution as if it were the true distribution, even if nothing is known about the convergence properties of the sequence of events. The justification is simple: If the sequence converges to a limit, then this rule will get us to the truth. First indications of this pragmatic approach, though no discussion, can be found in "Die Kausalstruktur der Welt".[84] In "Axiomatik der Wahrscheinlichkeitsrechnung",[85] Reichenbach presents the following axiom of induction as an additional axiom within a framework of inductive logic and as solution to the problem of ascertainability:

**Axiom of Induction:**
*If a physical description specifies an arbitrarily extendable sequence of frequency values* $h_i$*, of which* $h_1, \ldots, h_n$ *are determined, and if the numbers* $s_1$ *and* $s_2$ *are such that from a particular frequency value* $h_m$ $(1 \leq m \leq n)$ *on, all* $h_i$ $(m \leq i \leq n)$ *are bound by* $s_1$ *and* $s_2$*, then there is a probability* $w$ *such that the limit of the sequence is between* $s_1$ *and* $s_2$*;* $w$ *increases towards 1, if* $n - m$ *tends to infinity and all* $h_i$ $(m \leq i \leq n)$ *remain between these two bounds.*

Here Reichenbach postulates a second-order probability (and implicitly a hierarchy of probabilities) that is supposed to regulate convergence and provide justification. But there are several problems with this axiom: There are no quantifiers over $s_1$ and $s_2$. If they were universally quantified, i. e., if the axiom held for *any* $s_1$ and $s_2$, then a sequence satisfying the axiom is trivially guaranteed to converge. However, we are still faced with the problem of how we could *know* the axiom to be satisfied for a particular sequence, given that we have only seen a finite part. If the quantification is existential then there is no convergence guarantee, since the sequence could oscillate within the bounds; and even under existential quantification, it is unclear how the ascertainability problem is solved. In the example Reichenbach considers as illustration of the

83. Reichenbach (1949).
84. Reichenbach (1925), p. 145.
85. Reichenbach (1932a), p. 614.

use of this axiom[86] he argues that if we consider ever smaller intervals and ever larger number of trials, then we observe (and know from experience) that the empirical frequency distribution converges to a continuous function (recall the argument with strike ratios in his dissertation). But this begs the question. It does not solve the ascertainability problem. What justification do we have to assume convergence in the limit before we see it? An infinite regress of higher-order probabilities does not provide justification, since we do not know the value of any of these probabilities. The question remains why the straight rule is the best way to get to the truth.

In *The Theory of Probability* Reichenbach drops the hierarchy of higher order probabilities and presents the straight rule purely in terms of posits and its justification as a pragmatic matter.

### 6.4 *The Theory of Probability* (1935/1949)

In 1935, during his time in Istanbul, Reichenbach published his major work on probability: *The Theory of Probability – an Inquiry into the Logical and Mathematical Foundations of the Calculus of Probability*. He published it again in 1949, with major revisions, in English. This work is the most thorough account Reichenbach ever gave on probability. It is a frequentist account with a pragmatic approach in terms of posits and wagers filling in the gaps where limiting frequencies are unknown (the so-called "straight rule"). It is both technically and philosophically founded in his dissertation, addresses many of the dissertation's shortcomings and summarizes his ideas since its publication. In particular, it is a clear and explicit rejection of the Kantian component of Reichenbach's dissertation thesis.

By the time of publication of *The Theory of Probability* probability had penetrated into a large number of new areas of investigation. Physics had come to accept laws of statistical mechanics which contain probability as part of their lawfulness. Quantum physics undermined the whole idea of the existence of certainty as described in a Laplacian world. More than just a representation of our ignorance, there was now a suggestion that certain processes could in principle not be further reduced beyond probabilistic relations. One further aspect that perhaps was not so prevalent in the sciences at the time, but one that Reichenbach certainly considered important again, was the relation of probability and causality. Without Kant's synthetic a priori support, causation needed to be tied back into the general epistemological framework.

At the time Reichenbach wrote his dissertation, hardly any attempts at an axiomatization of probability theory were available and certainly none that were generally accepted (Kolmogorov published in 1933). The few attempts that were around were unknown to Reichenbach. In *The Theory of Probability* Reichenbach confronts this dearth of systematization by developing his own axiomatization of probability calculus (it was first published in a paper in 1932).[87] This

86. Reichenbach (1932a), p. 615.
87. Reichenbach (1932a).

axiomatization is similar to the well-known Kolmogorov axioms. It provides the foundations of a calculus independent of interpretation and a basis for Reichenbach to readdress with new rigor some of the technical aspects of the dissertation that had lacked precision.

As for the interpretation of probability, Reichenbach argued in the subsection of his dissertation on composite probability that the independence of trials is a necessary condition for a probability distribution, i. e., that the joint distribution of a sequence of events factors into the distributions over the indexed events. He argued that such a factorization could be obtained when the trials were causally independent, and hence pushed his requirement back to causally independent trials. Since Reichenbach had abandoned causality as a primitive notion and had found shortcomings in the requirements for randomness, he based his notion of probability in *The Theory of Probability* on normal sequences:

*A sequence B is said to be normal if it is free from after-effect and if the regular divisions belong to its domain of invariance.*[88]

A sequence is free of *after-effects* if the probability of a particular item occurring in the sequence is independent of the items that occurred before. A *regular division* is a subsequence selection rule that selects items from the original sequence in regular intervals, e. g., items 2, 6, 10, 14, etc. A subsequence is part of the *domain of invariance* if the event probabilities in the subsequence are the same as the event probabilities in the original sequence.

Reichenbach presents this notion of normal sequences in terms of (phase) probabilities. This is misleading as it alludes to a circular account of probability, where probabilities are defined in terms of normal sequences, which themselves can only be specified in terms of probabilities. However, the idea seems to be that normal sequences can be specified entirely in terms of particular types of subsequence selection rules and invariance conditions of limiting relative frequencies across different subsequences.

The probability of an event is then just the limiting relative frequency of the event in a normal sequence. This interpretation, Reichenbach claims, is based on work by Poisson, Boole and von Mises. His account allows him to consider a broad class of sequences amenable to probability claims. He avoids the circularity inherent in a definition of probability based on independent identically distributed trials. In fact, he is able to consider sequences of trials that are not entirely independent, as long as the relative frequencies converge and the sequence is normal. However, it places the real difficulty on the problem of ascertainability of probability claims.

Chapter 9 in *The Theory of Probability* can almost be seen as a de-Kanted version of Reichenbach's dissertation. He returns to the questions of meaning and ascertainability of probability claims. The problem, of course, is that under a

88. Reichenbach (1949), p. 144.

limiting frequency interpretation one can only make probability claims about infinite sequences that are described intensionally, but not about ones given extensionally. How, when observing a sequence of trials, can we know whether there is a limit, and if so, whether our current empirical distribution is representative of it? In his dissertation he suggested that a synthetic a priori principle ensures the existence of a (limiting) probability function, but Reichenbach realizes that this is unsatisfactory, since such a synthetic a priori claim is unverifiable. In footnote 1, paragraph 69 of *The Theory of Probability* Reichenbach acknowledges the Kantian influence in his dissertation and emphasizes that his views have thus changed. Reichenbach replaces the Kantian account with a pragmatic one: The practical limits of infinite sequences are good enough for our general purposes. We apply a probability calculus about infinite sequences on the basis of finite experience, just as we apply idealized geometrical laws in science to the real world that does not exactly satisfy those geometrical features.

The empirical distribution is our best bet at the truth. Basing our decisions on the empirical frequency distribution on the assumption that it more or less approximates the true distribution is the best we can do. Why? – Because if frequencies converge, we placed our bets well, if not, then there is no way of doing any better. The positive justification seems at least pragmatically plausible (although there may be doubts about the optimality), but Reichenbach does not spell out reasons for the latter negative justification. Furthermore, this account provides no basis for a measure of confidence in the empirical frequency distribution and so it is fair to say that Reichenbach does not really provide any justified insights to solving the problem of ascertainability.

In summary, *The Theory of Probability* provides two changes with regard to the dissertation: First, Reichenbach moves away from an account of probability based on causal independence of trials to one that is based on a form of near-random sequences formalized in the notion of *normal* sequences. Second, he drops the Kantian synthetic a priori guarantee of the existence of a probability function and replaces it with a frequentist interpretation of probability, where the empirical distribution is supposed to provide the "best" bet for an approximation of the true distribution.

## 7. Contemporary Echoes

Reichenbach's doctoral dissertation and the modifications of it that are developed in his posthumous *The Direction of Time* have echoes in contemporary philosophy and statistical science.

Reichenbach is arguably the first philosopher after Boole to inquire carefully into the connection between causality and probability. While Reichenbach is committed to a deterministic notion of token causation in his early work, including his dissertation, he soon develops these ideas into a theory of type-causation. His theory becomes most explicit in *The Direction of Time*, but its roots appear much earlier. Reichenbach's principle of the common cause cap-

tures the notion of screening off: If $C$ is a common cause of $A$ and $B$, then $A$ and $B$ are unconditionally correlated, but $A$ and $B$ are independent conditional on $C$. This principle is a special case of the causal Markov principle, first formulated by Kiiveri and Speed,[89] but fully used in the contemporary framework of causal Bayes nets and graphical causal models.[90] It links causal structure among variables with probability distributions. The principle underlies virtually all models of type-causation and has enabled a more thorough understanding of the relation between causality and probability. Arguably, historical justice would term it the Markov/Reichenbach Principle. Reichenbach also anticipates the connection between interventions, on the one hand, and causal and probabilistic relations on the other hand, a connection that has been developed into a widely cited formal tool for predicting the effects of interventions.[91]

One of the key problems underlying Reichenbach's analysis of probability is how the varying outcomes of individual events combine to form some fixed, nonvarying distribution which can be described by a probability claim. What are the features that have to be satisfied by the repeated process such that the variations in the individual trials cancel out to support one probability claim about the process? Or, in terms of the theory of error: Which conditions have to be satisfied in order to be able to identify the true effect from measurements that contain error? Reichenbach argues in his dissertation that any process which satisfies a certain continuity constraint about the distribution exhibited by the strike ratios will support probability claims. This argument has reappeared in the contemporary literature. In *Bigger than Chaos* (2003) Michael Strevens considers the problem of how complexity at the micro level can result in simple features at the macro level. He essentially restates the problem Reichenbach addresses of how variability at the micro level aggregates to form a stable probability at the macro level. His procedure and conditions are almost exactly Reichenbach's, but in a unique terminology.

Strevens offers a theorem that can be restated in something closer to Reichenbach's terms: If some designated outcome $e$ of an experiment exhibits a density to which strike ratios can be applied and if the empirical distribution approximates a Riemann integrable function, then the probability of $e$ is approximately equal to the strike ratio for $e$.

Strevens – like Reichenbach – cites Poincaré's work on the method of arbitrary functions. He points out that such an account has the advantage of not requiring any features of the limit distribution (the "IC-density" for Strevens) – no requirements *"other than that it exists"*. Nothing could resemble Reichenbach more closely.[92]

89. Kiiveri and Speed (1982).
90. Spirtes *et al.* (1993, 2000); Pearl (2000).
91. Spirtes *et al.* (1993, 2000); Pearl (2000); Woodward (2003).
92. It seems that Strevens was unaware of Reichenbach's dissertation. It is not cited and does not appear in the references. He refers to Reichenbach in the appendix of chapter 2 in a discussion of the method of arbitrary functions citing *The Theory of Probability*. He describes Reichenbach as combining Poincaré's frequency approach with an argument from

## 8. Epilogue

Reichenbach's doctoral dissertation looks backward, combining mathematical ideas borrowed from Poincaré and others with Reichenbach's understanding of the requirements of Kantian "objectivity". And yet the dissertation presages a lifetime of philosophical ideas that strike at fundamental issues that now range across all of the sciences: What are probabilities such that we can come to know them? What is causality and what is its logic? How are causality and probability related such that one is a guide to the other? No wonder, then, that Reichenbach's dissertation finds more echoes in contemporary philosophical and scientific discussions than do the inaugural works of other great 20th century thinkers, Bertrand Russell and Rudolf Carnap, for example, who, like Reichenbach, began their careers with efforts to reconcile Kantian philosophy with post-Kantian science. If, a century later, Reichenbach's arguments do not satisfy us, they nonetheless remind us vividly of our epistemological predicaments, and how we came to know them.

## Bibliography

Apelt, E. F. 1854. *Theorie der Induktion*. W. Engelmann.

Bernstein, F. 1907. "Über das Gaußsche Fehlergesetz" *Mathematische Annalen* 64, p. 417.

Borel, E. 1909. "Les probabilités dénombrables et leurs applications arithmétiques" *Rendiconti del Circolo Matematico di Palermo* 27, pp. 247–271.

Cassirer, E. 1910. *Substanzbegriff und Funktionsbegriff*. B. Cassirer.

Church, A. 1940. "On the concept of random sequence" *American Mathematical Society Bulletin* 46, pp. 130–135.

Copeland, A. 1928. "Admissible numbers in the theory of probability" *American Journal of Mathematics* 50, pp. 535–552.

Cournot, M. A. A. 1843. *Théorie des Chances et des Probabilités*. Hachette.

Czuber, E. 1891. *Theorie der Beobachtungsfehler*. B. G. Teubner.

———. 1908. *Wahrschcinlichkcitsrcchnung*. B. C. Tcubncr.

Elsas, A. 1889. "Kritische Betrachtungen über die Wahrscheinlichkeitsrechnung" *Philosophische Monatshefte*, p. 557.

Fick, A. 1883. *Philosophischer Versuch über die Wahrscheinlichkeiten*. Stahel.

symmetry. It is true that Reichenbach presents a frequentist account of probability, but he explicitly rejects symmetry arguments. As discussed above, Reichenbach is all too aware of the paradoxes he would get himself into on a symmetry account. Neither Strevens nor Reichenbach (in his dissertation) want to or do commit to a limiting frequency meaning of probability.

Gerner, K. 1997. *Hans Reichenbach, sein Leben und Wirken*. Phoebe Autorenpresse.

Grelling, K. 1910. "Die philosophischen Grundlagen der Wahrscheinlichkeitsrechnung" *Abhandlungen der Friesschen Schule III* 3.

Hausdorff, F. 1901. "Beiträge zur Wahrscheinlichkeitsrechnung" *Bericht der mathematisch-physischen Klasse der Königlich Sächsischen Gesellschaft der Wissenschaften* 53, pp. 152-178.

Hopf, E. 1934. "On causality, statistics and probability" *Journal of Mathematics and Physics* 13, pp. 51-102.

Kant, I. 1787. *Kritik der reinen Vernunft*. 2. Aufl. Akademieausgabe.

Kiiveri, H. and Speed, T. 1982. "Structural analysis of multivariate data: A review" In *Sociological Methodology*, S. Leinhardt, ed., San Francisco, CA: Jossey-Bass.

Lange, F. A. 1894. *Logische Studien*. Baedeker.

Laplace, P. S. 1840. *Essai Philosophique sur les Probabilités*. Bachelier.

Martin-Löf, P. 1970. "On the notion of randomness" In *Intuitionism and Proof Theory*, J. Myhill, A. Kino, and R. E. Vesley, eds., Amsterdam: North-Holland. pp. 227-234.

Nagel, E. 1938. "Principles of the theory of probability" In *Foundations of the Unity of Science*, R. Carnap, C. Morris, and O. Neurath, eds., Chicago, IL: University of Chicago Press.

Natorp, P. 1910. *Die logischen Grundlagen der exakten Wissenschaften*. B. G. Teubner.

Pearl, J. 2000. *Causality*. Oxford University Press.

Poincaré, H. 1912. *Calcul des Probabilités*. Gauthier-Villars.

——. 1914. *Wissenschaft und Methode*. B. G. Teubner.

Reichenbach, H. 1891-1953. "Unpublished notes" In *Reichenbach Collection*, Pittsburgh, PA: University of Pittsburgh Library System. All rights reserved. 1) ASP/HR 017-04 (notes on other authors for dissertation); 2) ASP/HR 025-04-01 (handwritten draft of dissertation); 3) ASP/HR 041-08-09 (letter to E. Lingener); 4) ASP/HR 044-06-21 (1927 comments on dissertation); 5) ASP/HR 044-06-22 (autobiographical notes, letter from Natorp, letter from G.E. Müller).

——. 1920a. "Philosophische Kritik der Wahrscheinlichkeitsrechnung" *Die Naturwissenschaften* 8, pp. 146-153.

——. 1920b. "Die physikalischen Voraussetzungen der Wahrscheinlichkeitsrechnung" *Die Naturwissenschaften* 8, pp. 46-55.

——. 1920c. *Relativitätstheorie und Erkenntnis a priori*. Springer.

——. 1925. "Die Kausalstruktur der Welt und der Unterschied von Vergangenheit und Zukunft" *Sitzungsberichte - Bayerische Akademie der Wissenschaften, mathematisch-naturwissenschaftliche Klasse*, pp. 133–175.

——. 1929. "Stetige Wahrscheinlichkeitsfolgen" *Zeitschrift für Physik* 53, pp. 274–307.

——. 1930. "Kausalität und Wahrscheinlichkeit" *Erkenntnis* 1, pp. 158–188.

——. 1932a. "Axiomatik der Wahrscheinlichkeitsrechnung" *Mathematische Zeitschrift* 34, pp. 568–619.

——. 1932b. "Die logischen Grundlagen des Wahrscheinlichkeitsbegriffs" *Erkenntnis* 3, pp. 410–425.

——. 1935. *Wahrscheinlichkeitslehre*. A. W. Sijthoff's uitgeversmij N.V.

——. 1936. "Warum ist die Anwendung der Induktionsregel für uns notwendige Bedingung zur Gewinnung von Voraussagen?" *Erkenntnis* 6, pp. 32–40.

——. 1938. "On probability and induction" *Philosophy of Science* 5, pp. 21–45.

——. 1940. "On the justification of induction" *The Journal of Philosophy* 37, pp. 97–103.

——. 1949. *The Theory of Probability*. University of California Press.

——. 1956. *The Direction of Time*. University of California Press. Edited by M. Reichenbach.

——. 1965. *The Theory of Relativity and A Priori Knowledge*. University of California Press. Edited by M. Reichenbach.

Salmon, W. C. 1979. *Hans Reichenbach: Logical Empiricist*. D. Reidel Publishing Company.

——. 1984. *Scientific Explanation and the Causal Structure of the World*. Princeton University Press.

——. 1998. *Causality and Explanation*. Oxford University Press.

Spirtes, P., Glymour, C., and Scheines, R. 1993, 2000. *Causation, Prediction and Search*. MIT Press, 2nd edn.

Strevens, M. 2003. *Bigger than Chaos*. Harvard University Press.

Stumpf, C. 1892a. "Über den Begriff der mathematischen Wahrscheinlichkeit" *Sitzungsbericht der philosophisch-historischen Klasse der königlich bayerischen Akademie der Wissenschaften zu München*.

——. 1892b. "Über die Anwendung des mathematischen Wahrscheinlichkeitsbegriffs auf Teile eines Continuums" *Sitzungsbericht der philosophisch-historischen Klasse der königlich bayerischen Akademie der Wissenschaften zu München*.

Suppes, P. 1970. *A Probabilistic Theory of Causality*. North-Holland.

Ville, J. A. 1936. "Calcul des probabilités - sur la notion de collectif" *Comptes Rendus Académie des Sciences* 203, pp. 26–27.

von Kries, J. 1886. *Die Prinzipien der Wahrscheinlichkeitsrechnung*. J. C. B. Mohr.

von Mises, R. 1919. "Grundlagen der Wahrscheinlichkeitsrechnung" *Mathematische Zeitschrift* 5, pp. 52–99.

Wald, A. 1938. "Die Widerspruchsfreiheit des Kollektivbegriffs" *Actualités Scientifiques et Industrielles* 735, pp. 79–99.

Woodward, J. 2003. *Making Things Happen*. Oxford University Press.

# Der Begriff der Wahrscheinlichkeit für die mathematische Darstellung der Wirklichkeit.

---

## Inaugural-Dissertation

zur

Erlangung der Doktorwürde
der hohen philosophischen Fakultät der Friedrich-Alexanders-Universität Erlangen

vorgelegt von

**HANS REICHENBACH**
aus Hamburg

Tag der mündlichen Prüfung: 2. März 1915

---

Leipzig 1916
JOHANN AMBROSIUS BARTH

# Die Bedeutung der Wahrscheinlichkeit für die mathematische Darstellung der Wirklichkeit

Hans Reichenbach

## 1 I. Das Problem

Noch immer hat der Streit um Grundbegriffe der Naturerkenntnis die Philosophen in zwei Lager geschieden. Nachdem die Grenze des primitiven Hinnehmens aller Erkenntnis als einer Selbstverständlichkeit überschritten worden war, ließen sich die einen durch die Entdeckung, daß in allen Urteilen subjektive Momente enthalten sind, dazu verleiten, den Glauben an eine objektive Erkenntnis überhaupt aufzugeben, und in aller Wissenschaft nichts anderes als ein Spiel der menschlichen Gedanken zu sehen. Die anderen jedoch ließen sich in ihrem natürlichen Vertrauen auf die objektive Gültigkeit wissenschaftlicher Resultate nicht erschüttern und stellten ihre Philosophie nicht so sehr auf die Frage ein, ob Erkenntnis wahr sei, als vielmehr, welches denn die notwendigen Elemente dieser Erkenntnis seien, die wir als wahr bezeichnen. Diese letztere Methode hat ihre systematische Ausbildung in der Kritik der Vernunft gefunden, die Kant geschaffen hat; und von ihrem Standpunkt muß das erstere Verfahren ebenso primitiv wie in sich widerspruchsvoll erscheinen.

Trotz der großen Entdeckung Kants ist jedoch eine kritische Untersuchung der in den positiven Wissenschaften angewandten Begriffe in nur geringem Umfange bisher durchgeführt worden. Nur wenige Philosophen sind den verzweigten Bahnen der Mathematiker und Physiker gefolgt und haben die Sonde der Kritik an die Methoden gelegt, die dort stetig angewandt werden; anstatt dessen hat man versucht, den Grundbegriffen, die gleichzeitig auch im täglichen Denken und Handeln des Menschen eine hervorragende Rolle spielen, von dieser
2 Seite der primitiven Betrach|tung her beizukommen und ihre Bedeutung aufzuklären. Das mußte zur Folge haben, daß wichtige Erkenntnisse, die in den Resultaten der positiven Wissenschaften bereits implizit vorliegen, der Philosophie vorerst verschlossen blieben, und daß andrerseits philosophische Begrif-

# The Concept of Probability in the Mathematical Representation of Reality

Hans Reichenbach

## 1. The Problem 1

The controversy surrounding the fundamental concepts of our knowledge of nature continues to divide philosophers into two camps. Since the primitive acceptance of all knowledge as self-evident has been questioned, one camp was led by the discovery that all judgments contain subjective elements to give up any belief in objective knowledge altogether and to view science as nothing else than a game played by human thought. The other camp, however, did not let their natural confidence in the objective validity of scientific results be shaken, and did not focus their philosophy on the question of whether our understanding is true, but rather on the elements that are necessary for the understanding that we deem true. This latter method found its systematic exposition in Kant's *Critique of Pure Reason*; and from its standpoint the former [subjective] method must seem both primitive and self-contradictory.

Despite Kant's great discovery, a critical analysis of the concepts used in the positive sciences has scarcely been done. Few philosophers have followed the winding paths of mathematicians and physicists and shone the light of criticism on the methods that are applied there; instead, philosophers have tried to grasp and clarify the meaning of these fundamental concepts – which also play a prominent role in everyday human thought and action – by means of primitive reflection. | This inevitably led to the consequence that important insights already 2
implicitly present in the positive sciences initially remained barred from philosophy and that philosophical concepts were created that bore no relation to the

fe geschaffen wurden, die zu den Bedürfnissen der Wissenschaft in keinerlei Verhältnis stehen. Insonderheit mußte dies bei einem Begriff hervortreten, der ebensowohl in der exakten Wissenschaft wie in der Praxis des täglichen Lebens einen wichtigen Platz einnimmt, dem Begriff der Wahrscheinlichkeit; hier mußten, da mit der Auffassung dieses Begriffes eine fundamentale Einstellung zum Wesen der Erkenntnis überhaupt gegeben war, die Differenzen um so weittragender werden. Die Mathematik kennt seit langer Zeit eine exakte Disziplin der Wahrscheinlichkeitsrechnung und hat mit großem Aufwand von Schärfe der Analyse und Präzision der Begriffe ein System von Sätzen über Wahrscheinlichkeitsbeziehungen geschaffen; wenn aber die philosophische Kritik die dort entdeckten Methoden unbenützt liegen ließ und sich allein mit dem etwas undeutlichen und stets verschwommen angewandten Wahrscheinlichkeitsbegriff des täglichen Lebens befaßte, wie er dort so häufig auftritt, so konnte sie zu einer präzisen Stellung des Problems, geschweige denn zu einer Lösung gewiß nicht kommen.

Um so mehr mußte diese Philosophie einem Rückfall in die Theorie des Subjektivismus ausgesetzt sein; denn die Art und Weise, wie im täglichen Leben Begriffe angewandt und Erkenntnisse gewonnen werden, ist gewiß nicht dazu geeignet, den Glauben an objektive Bedeutung solcher Aussagen zu wecken. Der Wahrscheinlichkeitsbegriff genießt da gewissermaßen noch eine Vorzugsstellung, denn wenn wir ein Geschehen wahrscheinlich nennen, so verzichten wir damit ausdrücklich auf die Behauptung der Gewißheit seines Eintretens, und man wird berechtigte Zweifel erheben, ob eine solche Behauptung irgendetwas Objektives über die Dinge aussagt. Hinzu kommt eine eigentümliche Gegensätzlichkeit zu dem allgemein als oberstes Prinzip des Geschehens annerkannten Kausalprinzip. Wir wissen, daß jedes wirkliche Geschehen notwendig bedingt ist, daß auch der zufällig vom Dache herabfallende Ziegelstein streng nach Gesetzen der Natur sich losgelöst hat, und daß der Fall schon vor jeder beliebigen
3 Zeit in den damaligen Zuständen der Dinge begründet ge|wesen ist; würden wir diese gekannt haben, so hätten wir den Fall sogar vorher als notwendig voraussagen können. Nur weil wir die spezielleren Zusammenhänge nicht bestimmen können, müssen wir uns bescheiden, den Fall als wahrscheinlich oder unwahrscheinlich anzunehmen; darf aber solche Aussage, die lediglich durch unsere Unwissenheit bedingt ist, den Anspruch auf objektive Geltung in irgendeinem Sinne erheben?

In der Tat haben sich viele Philosophen durch diese Paradoxie verleiten lassen, in dem Wahrscheinlichkeitsbegriff lediglich unsere subjektive Erwartung dargestellt zu sehen, die zur Welt der wirklichen Dinge keinerlei Beziehung hat. Ebenso ist die entgegengesetzte Ansicht entwickelt worden; doch ist es bezeichnend, daß auch diese das subjektive Moment aus dem Wahrscheinlichkeitsbegriff noch nicht restlos eliminiert hat; man kann sagen, daß je enger der betreffende Forscher sich an die mathematische Behandlung des Problems anlehnte, er um so mehr sich der Ansicht von der objektiven Bedeutung des Begriffs anschloß. Die gegensätzlichen Ansichten sind vornehmlich in zwei Arbeiten zum Ausdruck gekommen.

needs of science. This had to become particularly evident with one concept that plays a prominent role both in the exact sciences and in daily life, the concept of probability. Here the differences between philosophy and science had to be especially significant, since the understanding of this concept indicates a fundamental attitude towards the nature of knowledge in general. For a long time mathematics has contained an exact probability calculus and, through in-depth analysis and the use of precise concepts, has developed a system of propositions about relations of probability. But since philosophical criticism left these methods unused and only concerned itself with the somewhat unclear and loosely applied concept of probability as it is commonly employed in everyday life, it could never reach a precise account of the problem, let alone an answer.

Moreover this philosophy risked returning to the theory of subjectivism, since the way in which concepts are used and insights are won in everyday life is certainly not suited to spark belief in an objective meaning of such expressions. In this context, the concept of probability in some sense still enjoys a privileged position, since if we call an event probable then we thereby explicitly forego the claim of certainty of its occurrence. Justified doubts will be raised whether such a claim says anything objective about things at all. In addition there is a peculiar opposition to the causal principle that is generally regarded as the highest of all principles governing events. We know that every real event is necessarily caused, that the brick that accidentally falls from the roof became loose strictly following the laws of nature and that the state of affairs at any previous point in time already determined the fall; | if we had known these circumstances, we 3
could have predicted the necessity of the fall. Simply because we are unable to determine the exact relationships we have to be content to consider the fall as probable or improbable. But may such a claim demand objective validity in any sense if it is only due to our lack of knowledge?

Indeed, this paradox has led many philosophers to believe that the concept of probability only represents our subjective expectation, which does not have any connection to the real world. Similarly, the opposite view has been developed; but it is noteworthy that this view has not completely eliminated the subjective element from the concept of probability, either. One can say that the closer the particular researcher followed the mathematical treatment of this problem, the more he subscribed to the view of an objective meaning of this concept. The opposing views were most prominent in two pieces of work.

**I.**

Carl Stumpf hat in einer längeren Abhandlung den Standpunkt der subjektiven Wahrscheinlichkeit vertreten. Er definiert:[1] „Jede beliebige Urteilsmaterie nennen wir $\frac{n}{N}$ wahrscheinlich, wenn wir sie auffassen können als eines von $n$ Gliedern (günstigen Fällen) innerhalb einer Gesamtzahl von $N$ Gliedern (möglichen Fällen), von denen wir wissen, daß eines und nur eines wahr ist, dagegen schlechterdings nicht wissen, welches."

Er nennt solche Fälle gleich möglich, über deren Bevorzugung wir nichts ausmachen können, und verzichtet demnach auf jedes objektive Kriterium dafür. So hält er es für richtig, wenn wir bei einem Würfel die Wahrscheinlichkeit jeder Seite gleich $\frac{1}{6}$ ansetzen, weil wir nicht wissen, welche Seite darankommen wird; stellt sich später heraus, daß der Würfel falsch war, so werden wir dann
4 eine andere Wahrscheinlichkeit angeben, weil wir nun | mehr über den Würfel wissen. Stumpf verzichtet darauf, zu entscheiden, ob eine Wahrscheinlichkeit objektiv richtig oder unrichtig ist. Die erste Bestimmung war richtig, denn sie entsprach dem Stand unserer Kenntnisse; die spätere Feststellung, daß eine Seite des Würfels mit Blei hinterlegt ist, macht die alte Bestimmung nicht unrichtig, sondern eröffnet uns nur den Weg zu einer neuen. Es liegt nicht eine Korrektur, sondern eine Veränderung der Wahrscheinlichkeit vor, erklärt Stumpf. Und da Stumpf an keinerlei objektive Bedingungen seiner Wahrscheinlichkeit gebunden ist, kann er die definierte Maßzahl auf jede Urteilsmaterie anwenden. Es braucht weiter nichts erfüllt zu sein, als daß die möglichen Fälle die Glieder einer totalen Disjunktion vorstellen. Wenn wir z. B. einen neuen Kometen beobachten, so wissen wir, daß er sich entweder auf einer Hyperbel, oder auf einer Parabel, oder auf einer Ellipse, oder auf einem Kreis um die Sonne bewegen wird. Da wir aber noch nicht in der Lage sind, zu bestimmen, welche dieser Kurven der Komet wählen wird, so müssen wir nach Stumpf die Wahrscheinlichkeit jeder Kurve gleich $\frac{1}{4}$ setzen, denn die Disjunktion enthält vier Glieder.

Stumpf hat recht, wenn er sagt, daß dieser Wahrscheinlichkeitsbegriff keinerlei Voraussetzungen oder Überzeugungen hinsichtlich der objektiven Welt einschließt; und ebenso, wenn er zugibt, daß ein solches Wahrscheinlichkeitsurteil über die objektive Welt nichts aussagt. Es enthält wirklich nichts, als eine Darstellung des jeweiligen Standes unseres Wissens. Anstatt zu sagen: ich weiß, daß der Komet sich entweder auf einer Hyperbel, oder auf einer Parabel, oder auf einer Ellipse, oder auf einem Kreis bewegen wird, kann ich nach Stumpf auch sagen, die Wahrscheinlichkeit, daß der Komet sich auf einer bestimmten dieser Kurven bewegt ist $\frac{1}{4}$. Die Wahrscheinlichkeitsurteile stellen nach dieser Auffassung eine Art von stenographierter Enzyklopädie vor, denn sie enthalten nichts als eine Beschreibung unserer subjektiven Kenntnisse. Nur erscheint es zweifelhaft, ob diese Art der Darstellung unseres Wissens nützlich wäre, ob nicht besser die alte Methode, es in Sätzen ohne Bruchstriche niederzulegen, vorzuziehen wäre.

1. Stumpf, *Begriff der Wahrscheinlichkeit*, S. 48. Vgl. das Literaturverzeichnis am Schlusse.

## I.

In an extensive discussion Carl Stumpf argues for the subjective view of probability. He defines:[1] "We say that any event is $\frac{n}{N}$ probable if we can conceive of it as one of $n$ elements (favorable instances) among a total of $N$ elements (possible instances), of which we know that exactly one is true, but we do not know which."[E1]

He calls those instances equally possible whose chance of occurrence we cannot determine and thereby foregoes any kind of objective criterion. Consequently he considers it correct if we assign to each face of a die the same probability of $\frac{1}{6}$, because we do not know which face will come up; if it later turns out that the die was biased, then we will assign a different probability because we now know | more about the die. Stumpf forgoes any decision as to whether a probab- 4
ility is objectively correct or not. The first specification was correct, because it corresponded to our state of knowledge; the later discovery that one of the faces of the die is leaded does not make the initial specification incorrect, it merely opens the door to a new specification. Stumpf explains that this is not a case of correction, but rather a change of probability. And since Stumpf is not bound to any objective conditions for his probability he can apply the defined measure to any event space. No further conditions need to be satisfied other than that the possible events describe a complete disjunction. For example, if we observe a new comet, then we know that its trajectory around the sun is either a hyperbola, a parabola, an ellipse or a circle. But since we are unable to determine which of these trajectories the comet will take, according to Stumpf we have to assign to each a probability of $\frac{1}{4}$, since the disjunction consists of four elements.

Stumpf is right in saying that his concept of probability does not contain any conditions or convictions with respect to the objective world; and similarly, when he admits that such a judgment of probability does not make any claims about the objective world. It consists of no more than a representation of our current state of knowledge. Instead of saying: I know that the comet moves on a trajectory that is either a hyperbola, a parabola, an ellipse or a circle, I can say according to Stumpf that the probability that the comet's trajectory will be one of these curves is $\frac{1}{4}$. In this view, judgments of probability describe a type of shorthand encyclopedia because they contain no more than a description of our subjective knowledge. It appears doubtful, however, whether this type of description of our knowledge is useful, and it might be better to stick to the old method of describing what we believe in sentences without fractions.

1. Stumpf, *Begriff der Wahrscheinlichkeit*, p. 48. See bibliography at the end.

E1. When read literally, this is a rather strange description of a supposedly subjective frequency account of probability, but Reichenbach's German quote matches Stumpf's original.

Vor allem ist in diesem Verfahren ein Umstand nicht zum Ausdruck gekom-
5 men, der im allgemeinen mit dem Begriff „gleich | mögliche Fälle“ wiedergegeben wird. Die Definition des Maßes der Wahrscheinlichkeit, von der auch die Stumpfsche sonst nicht abweicht, unterscheidet sich doch in dem Punkte von der Stumpfschen, daß sie die sämtlichen möglichen Fälle als *gleich* möglich voraussetzt. So werden wir die Seiten eines falschen Würfels nicht als *gleich* mögliche Fälle betrachten, sondern nur die eines symmetrisch gebauten. Wir würden auch mit der Wahrscheinlichkeitsbestimmung $\frac{1}{4}$ für die hyperbolische Bahn des Kometen gewiß nicht zufrieden sein; denn die Parabel und der Kreis stellen nur Grenzfälle der Bahnkurve vor, und es ist offenbar sehr viel wahrscheinlicher, daß der Komet die Hyperbel oder die Ellipse wählt, als daß er gerade den ausgezeichneten Zwischenfall annimmt. Stumpf würde allerdings dies auch aus seiner Auffassung zu rechtfertigen versuchen. Er würde sagen, daß wir eben wissen, daß die parabolische Bahn nur bei einem einzigen Wert einer gewissen Bahnkonstanten erreicht wird, die hyperbolische jedoch bei sehr vielen Werten; und das unbefriedigende Gefühl rührt daher, daß dieses Wissen in dem Wahrscheinlichkeitsurteil noch nicht zum Ausdruck gekommen ist. Erst die kompliziertere Berechnung der Wahrscheinlichkeit, bei der der Ellipse und Hyperbel viel größere Zahlen zufallen als den beiden anderen Kurven, sei eine Darstellung der *Gesamtheit* unseres diesbezüglichen Wissens. Es ist jedoch nicht recht einzusehen, wie diese Erweiterung aus der Stumpfschen Definition folgt. Seine oben wiedergegebene Definition jedenfalls enthält nichts, woraus man die Gleichheit der möglichen Fälle bestimmen sollte. Das tritt nicht nur in diesem Beispiel, sondern allgemein bei Berechnung der sogenannten geometrischen Wahrscheinlichkeit hervor, wo die Wahrscheinlichkeit der Größe von Flächenstücken oder Raumstücken proportional wird. Die Vollständigkeit der logischen Disjunktion genügt hier nicht, sondern es kommt eine Messung objektiver Größen, von Raumgrößen, hinzu. In einer späteren Arbeit versucht Stumpf, seine Definition auch für diesen Fall zu rechtfertigen; doch ist es gewiß keine Lösung, wenn er schreibt: „Gleiche Unkenntnis ist in bezug auf die resultierende Wahrscheinlichkeit äquivalent mit Kenntnis der Gleichheit.“[2]

6 Übrigens hat Stumpf Unrecht, sich für seine Anschauung auf Laplace zu berufen. Zwar ist diesem großen Manne die ungeschickte Erklärung der gleich möglichen Fälle entschlüpft: «c'est à dire tels que nous soyons également indécis sur leur existence,» aber da Laplaces Arbeit, auch der „Essai philosophique,“ vorwiegend mathematisch orientiert ist, ist die philosophische Inkorrektheit zu entschuldigen. Daß in diesem Begriff der Gleichmöglichkeit das große Problem sich verbirgt, weiß Laplace sehr wohl, denn er schreibt:[3] «Mais cela suppose les divers cas également possibles. S'ils ne le sont pas, on déterminera d'abord leurs possibilités respectives dont la juste appréciation est un des points les plus délicats de la théorie des hasards.»

2. Stumpf, *Anwendung des Wahrscheinlichkeitsbegriffs*, S. 687.
3. Laplace, *Essai philosophique*, S. 12.

In particular this procedure does not take into account one aspect that is generally expressed by the concept | of "equally possible events". Nevertheless, 5
the definition of the measure of probability, which Stumpf's account otherwise retains, does differ from Stumpf's view in the point that all possible events are assumed to be *equally* possible. We do not consider the faces of a weighted die to be *equally* possible, but only those of a symmetric one. We would probably not be content with a probability of $\frac{1}{4}$ for each hyperbolic comet trajectory either, since the parabola and the circle are only limiting cases of the trajectory and it is obviously much more likely that the comet will move on a hyperbola or ellipse rather than following the special case. However, Stumpf would attempt to justify this within his account, too. He would say that we know that a parabolic trajectory only occurs for one particular value of a certain trajectory-constant, but that a hyperbolic trajectory results from several values of that constant. The uncomfortable feeling [about the value $\frac{1}{4}$] results from the fact that this knowledge had not yet been captured by the probability judgment. Only the complicated calculation of the probability that assigns larger numbers to the ellipse and hyperbola than to the other two curves would represent the *totality* of our relevant knowledge. But it remains unclear how this extension follows from Stumpf's definition. In any case his definition above contains nothing that allows one to determine the equality of possible events. This problem does not only arise in this example, but in general with calculations of so-called geometric probability where the probabilities are proportional to area or volume. The complete logical disjunction is insufficient, since a measurement of objective quantities, e. g., of volumes, has to be added. In a later work Stumpf attempts to justify his definition for this case, too, but he certainly provides no solution when he writes: "With respect to the resulting probability equal lack of knowledge is equivalent to knowledge of the equality."[2]

As an aside, Stumpf is wrong to attribute his view to Laplace. It is true 6
that this great man let slip a clumsy explanation of equi-possible events: "that is to say that we are equally uncertain of their existence", but since the principal focus of Laplace's work, including the "Essai philosophique", is mathematical, this philosophical incorrectness is to be excused. Laplace knows very well that the concept of equi-possibility contains this great problem, since he writes:[3] "But that supposes the various cases equally possible. If they are not so, we will determine first their respective possibilities, whose exact appreciation is one of the most delicate points of the theory of chances."[E1]

2. Stumpf, *Anwendung des Wahrscheinlichkeitsbegriffs*, p. 687.
3. Laplace, *Essai philosophique*, p. 12.

E1. Truescott and Emory's translation of Laplace's *Essai philosophique*, (New York: Dover, 1995) p. 11.

Der Stumpfsche Wahrscheinlichkeitsbegriff ist zwar in sich widerspruchslos, und kann deshalb nicht falsch genannt werden, da Definitionen willkürlich sind. Aber er leistet jedenfalls nicht das, was man allgemein von einem Wahrscheinlichkeitsbegriff verlangt. Denn er ist nicht geeignet, ein Maß der vernünftigen Erwartung abzugeben. Ein solches muß über die wirklichen Dinge etwas aussagen: während Stumpfs Wahrscheinlichkeitsbegriff lediglich eine Darstellung unseres Wissens enthält. Mag sein, daß unsere Erwartung sich stets nach dem Stand unseres Wissens regelt; die wirklichen Dinge aber tun es gewiß nicht. Die Wahrscheinlichkeitsrechnung aber hat nicht die Aufgabe, zu bestimmen, welches unsere Erwartung *tatsächlich* ist, sondern welches sie vernünftigerweise *sein soll.* Sie sucht nach einer Norm, sie fragt nach der richtigen Erwartung, die einen Aufschluß über die Zukunft der Dinge enthält, und die somit von einer objektiven Gesetzmäßigkeit entlehnt sein muß. Nach einer solchen müssen wir deshalb jetzt suchen, und sollte es uns nicht gelingen, eine zu finden, so müssen wir auf die Aufstellung einer Erwartung verzichten; nicht aber dürfen wir uns damit begnügen, zufällige Feststellungen unseres subjektiven Zustandes an ihre Stelle zu setzen.

## II.

Johannes von Kries hat die umfassendste und gründlichste Arbeit über das Problem geschrieben. Er ist den Wahrscheinlichkeitssätzen im einzelnen nachge-
7 gangen und hat den Inhalt dessen, | was mit den gleich möglichen Fällen gemeint ist, zu formulieren gesucht. Wir können seine Resultate in folgende Sätze zusammenfassen:

1. Wahrscheinlichkeitsurteile sind in bezug auf die wirklichen Dinge entweder wahr oder falsch.

2. Sie behaupten eine bestimmte, eigentümliche Struktur von der Wirklichkeit, für die Kries den Namen einer Struktur der „Spielräume des Verhaltens“ eingeführt hat.

3. Auf Grund dieser im besonderen Fall durch Wahrnehmung und Überlegung festzustellenden Struktur behaupten sie eine Gestaltung des zukünftigen Geschehens.

4. Dieser Behauptung liegt ein nicht empirisches Prinzip zugrunde, das „Prinzip der Spielräume“.

Zu 1. Stumpf berechnet die Wahrscheinlichkeit der Seite eines falschen Würfels, dessen Unsymmetrie aber unbekannt ist, zu $\frac{1}{6}$; wenn dann durch genauere Untersuchung die abweichende Lage des Schwerpunkts bekannt wird, so berechnet er zwar die veränderte Wahrscheinlichkeit, aber er nennt die frühere Berechnung nicht falsch. Es liegt für ihn nicht eine Korrektur, sondern eine Veränderung vor; für den früheren Stand unseres Wissens war die frühere Wahrscheinlichkeitszahl richtig, und ein objektives Richtig oder Falsch gibt es

It is true that Stumpf's concept of probability is free from self-contradiction, and can therefore not be called false, since definitions are arbitrary. But it does not succeed in achieving what is generally expected from the concept of probability. [Stumpf's account] is inadequate to specify a measurement of rational expectation. This requires a claim about the real world: Stumpf's concept of probability only expresses a representation of our knowledge. It may be that our expectation is always determined by our state of knowledge; but the real world is certainly not. However, the calculus of probability is not supposed to determine what our expectation *actually is*, but rather what it rationally *should be*. It seeks a standard and asks for the right expectation containing insights about the future, and therefore it has to be based on an objective regularity. We have to search for such a regularity, and if we should not succeed in finding it then we have to abandon the notion of expectation; but we should not be content instead with coincidental discoveries of our subjective state.

## II.

Johannes von Kries has written the most extensive and thorough work on this problem. He examined the theorems of probability in detail and attempted to
express | the meaning of equi-possible events. We can summarize his results as 7
follows:

1. Judgments of probability are with respect to the real world either true or false.

2. They describe a particular idiosyncratic structure of reality, which Kries called the structure of "event spaces".[E2]

3. On the basis of this structure, which can in special cases be determined by reason and perception, judgments can be made about the configuration of future events.

4. The basis of this claim is a non-empirical principle, the "principle of event spaces".

On 1. Stumpf determines the probability of a side of a biased die, whose bias is unknown, as $\frac{1}{6}$; if further investigation proves the center of gravity to be off-center, then he recomputes the probability, but he does not consider the initial calculation wrong. It is not a case of correction but one of change; on the basis of our earlier knowledge, the first probability was correct – there is no such thing as an objective right or wrong. Kries, on the other hand, considers the first determ-

E2. We translate "Spielräume" as "event spaces". In *The Theory of Probability* Reichenbach refers to "events" and "spaces" when he discusses in chapter 9 material closely resembling his thesis. However, in a reference to von Kries (p. 353) he leaves "Spielraumtheorie" untranslated. Other authors have translated "Spielraum" elsewhere as "field" or "range".

für ihn nicht. Kries dagegen nennt die erste Bestimmung falsch, die zweite richtig; damit bringt er zum Ausdruck, daß es ein objektives Maß gibt, welches in der Wahrscheinlichkeitszahl näherungsweise dargestellt werden soll.

Zu 2. Dieses objektive Verhältnis bringt Kries in dem Begriff des Spielraums zum Ausdruck. Ein Vorgang sei derart, daß verschiedene mögliche Erfolge an ihn geknüpft sind; die Gesamtheit der möglichen Wirkungen ist dann der „Spielraum des Verhaltens." Dieser Gesamtspielraum soll nun weiter teilbar sein in kleinere Bereiche, die Teilspielräume; diese sollen gleich groß sein, d. h. sie sollen sich in Zahlengrenzen gleichen Abstandes einschließen lassen. Die Meßbarkeit wird also vorausgesetzt. Beim Würfel z. B. wird es einen gewissen, zahlenmäßig meßbaren Bereich der Anfangsbedingungen geben, für den gerade das Auffallen einer bestimmten Seite als Erfolg erscheint; andere Anfangsbereiche führen zu anderen Seiten. Hier hätten wir sechs gleiche Teilspielräume
8 vor uns. Doch bedarf dies noch weiterer Bestim|mung. Es muß hinzukommen, daß keiner der Teilspielräume vor dem anderen bevorzugt ist; Kries erklärt dies so: die Aufstellung gleicher Teilspielräume ist nur dann zulässig, wenn unsere Kenntnis derart ist, daß sie zwar Werte außerhalb des Gesamtspielraums ausschließt,

> „aber durchaus keinen Grund enthält, innerhalb dieses Spielraumes einen Wert für wahrscheinlicher als irgendeinen anderen zu halten. Nur unter dieser Voraussetzung wird für die logische Berechtigung einzelner Annahmen ausschließlich die Größe der umfaßten Bezirke maßgebend sein; wir hätten es dann mit einer, in keinerlei Weise durch Gründe, sondern lediglich durch Größenverhältnisse geleiteten Abwägung von Annahmen zu tun. Wir wollen diese als *freie Erwartungsbildung* bezeichnen, und es mögen Spielräume des Verhaltens, für welche in der eben charakterisierten Weise keinerlei logische Bevorzugung des einen vor dem anderen besteht, *indifferent* genannt werden" (S. 25).

Hier hat Kries wieder von dem „Prinzip des mangelnden Grundes" Gebrauch machen müssen. Er trägt somit einen subjektiven Faktor in seinen Wahrscheinlichkeitsbegriff hinein, aber außer diesem läßt er andere, objektive Bestimmungen gelten.

Noch eine Bedingung müssen die Spielräume erfüllen. Man denke sich jeden Teilspielraum den Bereich der Bedingungen zugeordnet, der ihm zeitlich vorausgeht; dann ist es möglich, daß, wenn die Teilspielräume gleich groß sind, die korrespondierenden Spielräume verschieden sind. Dann aber könnte man die Teilspielräume nicht als gleichwahrscheinlich bezeichnen. Es muß deshalb in der Kette der Ursachen soweit zurückgegangen werden, bis Spielräume auftreten, die nicht weiter auf andere bezogen werden; diese sogenannten „ursprünglichen" Spielräume sind dann gleich groß zu machen, und die ihnen korrespondierenden Anfangsspielräume, die jetzt vielleicht ganz verschieden groß sind, sind dann gleichwahrscheinlich. Kries meint, daß solche ursprünglichen Spielräume z. B. bei Naturkonstanten gegeben seien; wenn das spezifische Ge-

ination [of probability] false and the second correct; he thereby commits himself to the existence of an objective measure that is represented approximately by a particular probability value.

On 2. For Kries, this objective relation is captured by the concept of an event space. Let a process be such that it has various outcomes; the totality of possible outcomes is then called the "universe of events." This universe of events is divisible into smaller parts, the event spaces; these are of equal size, i. e., they should be contained within equidistant numerical boundaries. So he must assume that measurement is possible. For example, in the case of a die there is a particular numerically measurable range of initial conditions that results in a particular face coming up; different initial conditions lead to different sides coming up. So here we have six equal event spaces. But further analysis is
required. | In addition, no event space may be more likely than another. Kries ex- 8
plains this in the following way: the specification of event spaces is only admissible when our knowledge is such that we can exclude values outside the universe of events,

> but we have no reason to consider one value within the universe of events more likely than any other. Only under this condition will the individual suppositions be logically justified solely by the size of the covered areas. We would thereby arrive at an assessment of the suppositions guided not by reason, but purely by proportions of sizes. Let this be called a *free formation of expectation* and let the events be called *indifferent* if none are logically more likely than others in the sense just described. (p. 25)

Here Kries again has to make use of the "principle of insufficient reason." He thereby introduces a subjective component into his concept of probability, but apart from this objective considerations apply.

Event spaces must satisfy one further condition. Suppose that every event were assigned to the set of conditions that is temporally prior to it; then it is possible that, if the events are of equal size, the corresponding spaces [of the conditions] are of different sizes. However, then the events [with which we began] cannot be considered equi-probable. One therefore has to trace the chain of causes until one finds events that do not draw upon other events; these so-called "elementary" event spaces are then to be made equal-sized and the events we considered initially, corresponding to these spaces of elementary events, may now have quite different sizes but are then equi-probable. Kries claims that such elementary events can, for example, be found in the context of natural constants; if the specific gravity of a substance lies between 5 and 6, then I can consider

wicht einer Substanz zwischen 5 und 6 liegt, so kann ich darin alle gleich großen Teilbereiche als ursprünglich ansehen, weil sich keine Ursachen angeben lassen, die der Naturkonstanten gerade den speziellen Wert geben. Diese Ansicht, daß die Naturkonstanten sozusagen ursachlos, nicht als Funktion darstellbar sind, ist jedoch falsch und wird später in anderem Zusammenhang widerlegt werden.
9 Nur der andere, von | ihm genannte Fall kann hier bestehen bleiben. Zwar hat jedes Geschehen seine Ursache, und so muß auch jeder Schar von Spielräumen eine frühere korrespondieren; aber es ist möglich, daß sich schließlich ein Zustand herstellt, von dem ab das Größenverhältnis der Spielräume nicht mehr geändert wird, wie weit ich auch zurückgehe. In diesem Fall nennt Kries die Spielräume ebenfalls ursprünglich.

Nach Kries können wir ein Wahrscheinlichkeitsmaß dann aufstellen, wenn wir ein Geschehen als in gleiche, indifferente und ursprüngliche Spielräume zerfallend darstellen können. Das Maß der Wahrscheinlichkeit ergibt sich dann als das Verhältnis aus der Anzahl der für den Erfolg günstigen Teilspielräume und ihrer Gesamtzahl. Ob dies erfüllt ist, darüber müssen im einzelnen Fall Erfahrung und Überlegung entscheiden. Umgekehrt, wird ein Wahrscheinlichkeitsmaß für einen bestimmten Vorgang aufgestellt, so behauptet es eine solche Struktur.

Zu 3. In dieser Behauptung liegt eine Aussage über das zukünftige Geschehen. Wir dürfen mit Recht erwarten, daß bei häufiger Wiederholung des Vorganges jeder Erfolg nach der ihm zugehörigen Wahrscheinlichkeit darankommen wird. Es erscheint jedoch bei Kries nicht klar, welcher Art die Behauptung über das zukünftige Geschehen ist. Daß wir mit Gewißheit eine derartige Verteilung behaupten können, bestreitet er, und das muß er auch, weil sein Wahrscheinlichkeitsbegriff noch ein subjektives Moment im Begriff der Indifferenz enthält. So wird es aber fraglich, ob sein Wahrscheinlichkeitsurteil überhaupt noch etwas von der zukünftigen Wirklichkeit aussagt.

Zu 4.

> „Das Prinzip der Spielräume... kann durch die Erfahrung weder bewiesen noch widerlegt werden. Zwar mag die Zuversicht, mit welcher wir es zur Anwendung bringen, nach psychologischen Gesetzen wachsen, wenn die Erfahrung die auf das Prinzip gebauten Erwartungen bestätigt. Gleichwohl kann von einem empirischen Beweise nicht die Rede sein, schon weil das Prinzip gar keinen Inhalt hat, der, irgendwelchen Erfahrungssätzen gleichartig, aus ihnen gefolgert werden könnte. Eine Nichtbestätigung der Erwartungen andererseits wird, gerade wie beim Kausalgesetz, zunächst die Forderung ergeben, daß die besonderen Annahmen empirischen Inhalts, nach Maßgabe deren das Prinzip zur Anwendung gebracht
> 10 wurde, unrichtig waren, d. h. daß die | Größenbeziehungen gewisser Verhaltungsspielräume andere waren, als wir geglaubt hatten.... Sollte aber die Nichtbestätigung unserer Erwartungen selbst da stattfinden, wo die Richtigkeit jener Bestimmungen über jeden

all equally sized partitions contained in this interval as elementary, because no causes can be specified that give the natural constant any particular value. This view, that natural constants are, as it were, free from causes and not representable as a function, is mistaken and will be refuted later in a different context. Only the other case | he mentions can be maintained here. Although every event 9
has a cause and consequently every set of events corresponds to an earlier one, it is nevertheless possible that a circumstance arises in which the relative sizes of events do not change no matter how far back one goes. In this case, also, Kries calls the event spaces elementary.

According to Kries we can develop a measure of probability if we are able to represent events as consisting of equal, indifferent and elementary event spaces. The measure of probability is then formed from the proportion of the number of successes over the total number of events. Whether or not these conditions are satisfied has to be determined in each case by experience and reason. Conversely, whenever a measure of probability is determined for a particular process, then it specifies such a structure.

On 3. This statement makes a claim about future events. We are justified in assuming that in the case of many repetitions of the process, each success will occur according to its corresponding probability. However, Kries is not clear on what type of claim this makes about future events. He denies that we can make such claims about the distribution with any certainty; he must, because his concept of probability still contains a subjective component in the notion of indifference. However, it then is doubtful whether his judgment of probability says anything at all about future reality.

On 4.

> Experience cannot prove or disprove ... the principle of event spaces. Indeed, confidence in applying it may increase according to psychological laws when experience confirms the expectations built on this principle. Still, one cannot speak of an empirical proof – the principle contains no content that can be inferred – as would be the case with claims of experience. On the other hand, if an expectation is not confirmed, this will result – as in the case of the causal law[E2] – in the claim that the particular empirical assumptions that led to the application of the principle were wrong, i. e., that the | relative sizes of 10
> particular event spaces were different than initially thought. ... If, however, our expectation fails to be confirmed in a case where there is no doubt as to the correctness of these assumptions then we would

E2. Von Kries is referring to Kant's principle of causality.

Zweifel erhaben wäre, so würden wir 'einen sehr merkwürdigen Zufall' statuieren; es ergäbe sich aber daraus keinerlei Konsequenz, welche die weitere Anwendung des Prinzips verhindern oder etwas anderes an seine Stelle setzen könnte." (S. 170)

Neben den hier angeführten vier positiven Resultaten müssen der Kriesschen Theorie aber zwei Mängel nachgesagt werden.

1. Die Formulierung des Prinzips als „Prinzip der Spielräume" erscheint noch recht unzureichend. Die drei Bestimmungsstücke gleich, indifferent und ursprünglich müssen sich sicherlich einfacher zum Ausdruck bringen lassen durch eine Formulierung, die sich enger an die mathematische Behandlung der Theorie anschließt als es die Kriessche Abhandlung tut. Übrigens sei hier hingewiesen auf eine Bemerkung Grellings, der dem Prinzip der Spielräume noch das der „gleichmäßigen Dichte" hinzufügt; dieses enthält eine wertvolle Ergänzung des Prinzips und bedeutet einen Schritt auf dem Wege zur schärferen Formulierung.

2. Wesentlich aber muß der Kriesschen Arbeit vorgeworfen werden, daß sie keinerlei Versuch macht, das einmal gewonnene Prinzip in das System der Philosophie einzugliedern. Wenn ein solches Prinzip der Forschung sich nachweisen läßt, das nicht empirisch bestätigt werden kann, so darf es doch nicht schlechtweg als vorhanden hingenommen werden, sondern die Rechtmäßigkeit seines Gebrauchs muß zuvor noch untersucht werden. Kries aber bestreitet die Notwendigkeit einer solchen Kritik. Er führt aus:

„Die Ähnlichkeit der logischen Bedeutung, welche das Prinzip der Gesetzmäßigkeit einerseits und das Prinzip der Spielräume andererseits hierbei ungeachtet ihrer sonstigen Verschiedenheit aufweisen, besteht offenbar darin, daß sie beide definitive, keiner weiteren Begründung oder Erklärung fähige Prinzipien sind, welchen gemäß wir Erwartungen bezüglich dieses oder jenes realen Verhaltens bilden. Trotz des Umstandes, daß wir dem Prinzip der Gesetzmäßigkeit eine objektive, dem Prinzip der Spielräume eine lediglich subjektive Bedeutung zuschreiben müssen, kommen sie doch hier beide in gleicher Weise in Betracht, denn auch jenes ist zugleich ein für die subjektive Gewißheit
11 be|stimmendes Prinzip und hat insofern eine ganz ähnliche Bedeutung wie das Prinzip der Spielräume. Mit welchem Rechte wir annehmen, daß gewisse Umstände gewisse Folgen notwendig herbeiführen, und mit welchem Rechte wir demgemäß erwarten, daß bei Realisierung solcher Umstände das, was ihre gesetzmäßige Folge ist, nun auch wirklich eintreten werde, das läßt sich ebenso wenig angeben, als mit welchem Rechte wir innerhalb eines großen Spielraums von Möglichkeiten die Verwirklichung einer ganz besonderen Form des Verhaltens für sehr unwahrscheinlich halten." (S. 160.)

> consider this to be a 'very curious coincidence', but there would not result any consequence that would prevent further application of the principle or replace it with something else. (p. 170)

In addition to these four positive results, two shortcomings of Kries's theory have to be mentioned.

1. The formulation of the "Principle of Event Spaces" seems insufficient. The three criteria – equal, indifferent and elementary – can surely be expressed more easily in a formulation that is closer to the mathematical treatment of the theory than is Kries's account. By the way, let me mention Grelling's remark which adds to the principle of event spaces the principle of "uniform density"; it contains a valuable supplement to the principle and is a first step towards a more precise formulation.

2. More importantly, however, Kries stands accused of failing to integrate his principle into the system of philosophy. If such a principle of research, which cannot be empirically confirmed, is discovered, then it cannot simply be assumed as given, but rather the correctness of its application has to be analyzed. Kries denies the need for such critical analysis. He says:

   > The principle of lawfulness on the one hand and the principle of event spaces on the other are similar in their logical meanings irrespective of their other differences. This similarity consists in the fact that they both are definitive principles that do not allow for any further justification or explanation and according to which we form expectations with respect to particular real behavior. Similar considerations apply to both principles even though we have to ascribe an objective meaning to the principle of lawfulness, but only a subjective one to the principle of event spaces, since the principle of lawfulness is at the same time a constitutive principle of subjective certainty | and in that sense it has a very similar 11
   > meaning to the principle of event spaces. We cannot specify the justification for our assumption that particular circumstances necessarily lead to particular consequences nor the justification for our expectation that in the case of the realization of these circumstances, the consequences actually occur according to the law. Similarly we cannot specify the justification for the fact that we consider the realization of one particular event in a very large space of possible events to be very unlikely. (p. 160)

Dies ist ein Irrtum. Die Rechtmäßigkeit der Anwendung des Kausalprinzips ist von Kant in der transzendentalen Deduktion der Kritik dargetan worden; wäre dies nicht möglich gewesen, so hätten wir nicht das Recht, aus diesem Prinzip eine subjektive Gewißheit zu entnehmen. Es geht nicht an, unsere Annahmen über die Natur regelnde „subjektive" Prinzipien aufzustellen, die nicht in objektiven Gesetzen des Geschehens ihren Grund haben. Wenn das Prinzip der Spielräume für unsere subjektive Erwartung Bedeutung haben soll, so muß es sich als ein objektives Gesetz der Natur rechtfertigen lassen. Die Unklarheit, die wir in der Bemerkung zu Satz 3 Kries' Anschauungen vorwerfen mußten, beruht wesentlich darin, daß Kries auf diesen Punkt nicht eingegangen ist. Die Untersuchung dieser Grundfrage des Problems, auf die bisher auch von anderer Seite keine Antwort gegeben worden ist, wird deshalb wesentliche Aufgabe dieser Arbeit sein.

## III.

Verschiedene andere Arbeiten (F. A. Lange, E. F. Apelt, A. Fick, K. Grelling) stimmen darin überein, daß sie das Wahrscheinlichkeitsurteil als disjunktives Urteil auffassen. Insbesondere hat A. Fick diesen Gedanken durchgeführt. Nach ihm läßt sich jedes Wahrscheinlichkeitsurteil als hypothetisches Urteil formulieren, dessen Vordersatz den Gesamtbereich einer Bedingung, dessen Nachsatz nur eine an einen Teil dieses Bereichs geknüpfte Folge darstellt. Das Größenverhältnis dieser Bereiche stellt dann die Wahrscheinlichkeit dar. Nach Fick kommt die Wahrscheinlichkeit nicht dem Ereignisse, sondern dem Urteil zu. Vielleicht ist es nicht unzweckmäßig, die Wahrscheinlichkeitsaussage in dieser Form den
12 Urteilen einzuordnen. Wenn aber nicht mehr von einer | Wahrscheinlichkeit des Ereignisses gesprochen werden soll, welches ist dann der Sinn dieser mit dem Namen Wahrscheinlichkeit bezeichneten Eigenschaft des Urteils? Was sagt dieses Urteil über das Geschehen der *Dinge* aus? Darauf wird durch die Einordnung des Wahrscheinlichkeitsurteils unter die disjunktiven Urteile keine Antwort erteilt.

Wenn man von der Beziehung der Wahrscheinlichkeitsurteile auf die wirklichen Dinge absieht, so lassen sie sich allerdings als rein mathematische Sätze behandeln. Die sämtlichen Probleme der Wahrscheinlichkeitsrechnung sind solche Sätze. Wenn die Wahrscheinlichkeit, mit einem Würfel 3 zu werfen, $\frac{1}{6}$ beträgt, wie groß ist dann die Wahrscheinlichkeit, mit zwei Würfeln 4 zu werfen? Dies ist in der Tat ein rein mathematisches Problem, und seine Lösung, die wieder in hypothetischer Form dargestellt werden muß, ist dann mit *Gewißheit* zu bestimmen. Weit verzweigte Sätze dieser Art sind abgeleitet worden und haben zur Bildung einer besonderen mathematischen Disziplin, der Wahrscheinlichkeitsrechnung, geführt. Hierher gehört auch das Bernoullische Theorem, das über das Wachsen der Wahrscheinlichkeit mit großen Wiederholungszahlen etwas aussagt. Es beantwortet lediglich die Frage: Wie verändern sich mit wachsender Zahl die mathematischen Größenverhältnisse, die man Wahrscheinlich-

This is a mistake. The correctness of the use of the principle of causality was shown by Kant in the transcendental deduction in the Critique; if this had not been possible, then we would have no right to base a subjective certainty on this principle. It cannot be that we form suppositions on the basis of "subjective" principles that control nature but are not founded in objective laws of events. If the principle of event spaces is to have any meaning for our subjective expectation, then it has to be justifiable as an objective law of nature. The ambiguity that we established in comment 3 on Kries's account is primarily due to the fact that Kries did not address this issue. The analysis of this fundamental issue, which so far remains unanswered, will be the primary focus of this work.

## III.

A variety of other works (F. A. Lange, E. F. Apelt, A. Fick, K. Grelling) agree in taking the probability judgment to be a disjunction. A. Fick, in particular, followed this line of thought. According to him, every probability judgment can be formulated as a hypothetical judgment whose antecedent represents the whole space of a condition, while its consequent only represents a subspace of this space. The ratio of these two spaces is then the probability. According to Fick, the probability does not refer to the event, but to the judgment. Perhaps it is not inconvenient to associate the probability claims with the judgments in this manner. But if one is not to speak of | probabilities of events, what then is the 12
meaning of this property of judgments that we call probability? What claim does this judgment make about *events*? Classifying the probability judgment as disjunctive does not answer this question.

Ignoring the relationship between judgments of probability and reality, judgments of probability can be treated as purely mathematical propositions. All problems of the probability calculus are such propositions. If the probability of throwing a 3 with a die is $\frac{1}{6}$, what is then the probability of throwing a 4 with two dice? Indeed this is a purely mathematical problem, and its solution, which again has to be represented in a hypothetical form, can then be determined with *certainty*. Complex propositions of this type have been derived and have led to the development of a special mathematical discipline, probability calculus. This includes Bernoulli's Theorem, which makes claims about the increase of probability for large numbers of repetitions. It merely answers the question: How do the mathematical ratios of sizes, called probabilities, change with increasing num-

keit nennt? Fick hat Recht, wenn er alle diese Sätze wie die mathematischen Sätze synthetische Urteile a priori nennt. In dieser Auffassung ist ihm Grelling gefolgt.

Diese mathematischen Sätze sind durch die Erfahrung nicht zu bestätigen; ob sie aber auf die Wirklichkeit anwendbar sind, ist ein besonderes Problem, daß jenseits der Disziplin der Wahrscheinlichkeitsrechnung durch philosophische Untersuchung gelöst werden muß. Unter allen Autoren hat meines Wissens allein Fick diese Fragestellung getroffen, wenn er seine Abhandlung mit den Worten beschließt: „Wenn ich aus dem Gefässe zehnmal ziehe und es erscheint wirklich fünfmal weiß, so ist dies ebensowenig eine Bestätigung jenes Satzes, wie es eine Widerlegung desselben ist, wenn einmal bei zehn wirklichen aufeinanderfolgenden Zügen keinmal weiß erscheint. Denn der Satz sagt nicht etwa aus, beim wirklichen Ziehen muß in der Hälfte der Fälle weiß erscheinen, sondern er
13 sagt eben nur und weiter gar | nichts aus als: die allgemeine Bedingung, ‘wenn ich aus diesem Gefäße ziehe,’ zerfällt in zwei gleiche Sphären, deren eine das Erscheinen von weiß, deren andere das Erscheinen von schwarz zur Folge hat. Daß gleichwohl zwischen der Wahrscheinlichkeit und der Wirklichkeit eine gleichsam asymptotische Beziehung stattzufinden scheint, stellt der Metaphysik ein Problem, welches vielleicht zu den lösbaren gehört.“

Diesem Problem sollen die folgenden Untersuchungen gelten. Dabei werden folgende methodischen Gesichtspunkte maßgebend sein.

Es ist eine psychologische Tatsache, daß wir täglich so handeln, als ob dem Wahrscheinlichkeitsgesetz Geltung in der Wirklichkeit zukomme. Wenn wir mit einem Würfel werfen, so zweifelt niemand daran, daß nach einiger Zeit jede Seite einmal darankommen wird. Aber nicht nur in den Zufallsspielen, sondern in wichtigen Fragen des Lebens und in wissenschaftlichen Dingen setzen wir tatsächlich die Gültigkeit der Wahrscheinlichkeitsgesetze voraus. In den modernen Versicherungsgesellschaften haben wir Institutionen, deren Erfolge nur dadurch möglich sind, daß Wahrscheinlichkeitsgesetze das wirkliche Geschehen regieren; und jeder Aktionär einer Versicherungsgesellschaft beweist durch den Ankauf seiner Aktien, daß er unbedingt auf diese Gesetze vertraut. Die moderne Physik ist längst über jeden Zweifel an der Anwendbarkeit von Wahrscheinlichkeitsgesetzen hinausgeschritten, sie hat in den Gebieten der Molekulartheorie, der Quantentheorie wichtige Grundgesetze der Natur durch Anwendung statistischer Betrachtungen aufgedeckt. All das gibt uns mit Recht den Anlaß, zu vermuten, daß in den Wahrscheinlichkeitsgesetzen objektive Gesetze des Naturgeschehens vorliegen, deren Geltung sich philosophisch begründen lassen muß.

Wir wollen deshalb zunächst einmal so tun, als ob ihre Geltung bereits bewiesen sei, und, von dieser Annahme ausgehend, aufzudecken versuchen, welches eigentümliche Verhalten der Natur wir damit voraussetzen. Die Kriessche Arbeit, welche in dieser Richtung liegt, kann uns dennoch nicht genügende Vorarbeit sein, weil wir nach einer exakteren Formulierung des Wahrscheinlichkeitsprinzips suchen müssen, als Kries sie in dem Prinzip der Spielräume gibt; insbesondere werden wir bemüht sein, das Prinzip vom mangelnden Grunde, das auch Kries nicht umgehen konnte und das als lediglich subjektiver Faktor

bers? Fick is right in calling all these propositions synthetic a priori judgments, just as are mathematical propositions. Grelling follows him on this point.

These mathematical propositions cannot be confirmed by experience; whether they can be applied to reality is a separate problem which must be solved by philosophical analysis, independent of the discipline of the probability calculus. To my knowledge, among all authors, only Fick hit upon this problem when he concludes his essay with the words: "If I draw ten times from the bowl and white in fact occurs five times, then this is just as little a confirmation of the proposition as it would be a refutation of the same if I happened to find no whites in ten consecutive draws. The proposition does not claim that white has to appear half the time in actual draws. It claims no more than this: | the general condition, 13
'if I draw from this bowl,' has two equal ranges, one results in the appearance of white, the other in the appearance of black. The fact that there appears to be an asymptotic relationship between probability and reality poses a problem for metaphysics which might be solvable."

Subsequent analyses are devoted to this problem. The following methodological considerations will form the guiding principles.

It is a psychological fact that we act in everyday life as if the laws of probability are applicable to reality. When we throw a die, no one doubts that each side will eventually appear. We assume the validity of the laws of probability not only in games of chance, but also in scientific considerations and in important questions of our life. With modern insurance companies we have institutions whose success is only possible because the laws of probability in fact govern real events, and by buying shares every shareholder of an insurance company shows that he fully trusts these laws. Modern physics has long transcended every doubt about the applicability of laws of probability, it has discovered important fundamental laws of nature by the application of statistical methods in the area of molecular theory and quantum theory. All this gives us occasion to suspect that the laws of probability are in fact objective laws of nature and that a philosophical justification of their validity must be possible.

We will therefore initially assume that their validity has been proven, and starting from this assumption we will attempt to discover which particular features of nature we are presupposing. Kries's work, which heads in this direction, will be insufficient preparation, because we must find a more exact formulation of the principle of probability than the one Kries provides in the principle of event spaces. In particular, we will strive to get rid of the principle of insufficient reason, which Kries could not avoid and which, since it is purely subject-

14 eine objektive Geltung | des Wahrscheinlichkeitsgesetzes ausschließen würde, auszuschalten. Wir werden uns deshalb eng an die mathematische Behandlung der Probleme anschließen. Es gibt keine bessere Vorarbeit für die Behandlung naturphilosophischer Probleme als die exakte Analyse, wie sie in den Arbeiten der mathematischen Physik und angewandten Mathematik niedergelegt ist. Der Mathematiker setzt all jene Prinzipien fortwährend voraus, deren Kritik Aufgabe des Philosophen ist, und indem er, der natürlichen Einsicht folgend, ihre speziellen Gesetze bestimmt und weiterführt, bringt er sie selbst erst ans rechte Licht und weist dem Philosophen den Weg, den die Vernunftkritik zu gehen hat. Wir werden deshalb zuerst einen ganz speziellen Fall eines physikalischen Wahrscheinlichkeitsproblems, der besonders instruktiv erscheint, exakt mathematisch behandeln, und dann zu dem komplizierten mathematischen Ausbau fortschreiten, den dieselben Grundgesetze in der Fehlertheorie gefunden haben. Erst dann werden wir die Voraussetzung, die wir am Anfang machten und deren Inhalt wir inzwischen scharf formulieren könnten, wieder fallen lassen, um in einem weiteren, philosophischen Kapitel sie selbst zum Gegenstand der Erörterung zu machen.

## 15 2. Analyse spezieller Wahrscheinlichkeitsprobleme

### 2.1 Die Wahrscheinlichkeitsmachine

Wir wollen uns eine ideale Vorrichtung konstruieren, die so beschaffen ist, daß nach allgemeinen vernünftigen Erwägungen jedermann auf sie die Gesetze der Wahrscheinlichkeitsverteilung anwenden würde. Wir werden dann, stets der natürlichen Einsicht vertrauend, zu analysieren versuchen, welche Momente es waren, die uns zur Annahme dieser Verteilung geführt haben. Obgleich wir zunächst nur die Bedingungen dieses besonderen Falles untersuchen, werden wir zu allgemeinen Resultaten gelangen; denn wir werden dabei auf die wesentlichen Grundlagen des Wahrscheinlichkeitsproblems überhaupt geführt werden und erkennen, daß uns das Beispiel nicht mehr gewesen ist als eine Veranschaulichung und willkommene Führung auf dem Wege zu allgemeinen Gesetzen. Es übernimmt etwa die Rolle der gezeichneten Figur bei einem geometrischen Beweis. Wir denken uns folgende Maschine:

Über zwei feste Rollen sei ein in sich geschlossenes, horizontales Band geführt, das mit schmalen, gleich breiten schwarzen und weißen Querstreifen bedeckt ist. Es wird durch die Rollen mit gleichförmiger Geschwindigkeit bewegt. In einiger Entfernung darüber ist ein senkrechter Zylinder aufgestellt, mit der Öffnung nach unten; in ihm ist ein Kolben beweglich, der durch eine Feder mit starkem Druck abwärts gepreßt wird. Vor ihm ist ein Geschoß lose befestigt.
16 Wird der Kolben, nachdem er nach oben | angezogen und die Feder gespannt ist, losgelassen, so schnellt er sehr rasch nach unten und schleudert das Geschoß abwärts; dieses trifft den rasch dahingleitenden Papierstreifen und durchbohrt ihn. Es möge ferner so eingerichtet werden, daß jedesmal zu Beginn eines Schusses sich gerade der rechte Grenzstrich eines weißen Querstreifens an der Stelle

ive, would preclude the objective validity | of the laws of probability. We will 14
therefore remain close to the mathematical treatment of the problem. There is no better preparation for the treatment of natural-philosophical problems than an exact analysis as it is done in studies in mathematical physics and applied mathematics. The mathematician continually assumes principles that should be criticized by the philosopher. By determining and developing the specific laws according to his natural intuition, he illuminates the underlying principles and indicates which path the philosopher should follow in his rational critique. Initially therefore we will consider with mathematical exactness a very specific case of a physical probability problem that appears to be very instructive. We then proceed to a more complex mathematical extension which the same fundamental principles realize in the theory of error. Only then will we be able to drop the assumptions we made at the beginning, whose content we will now be able to formulate precisely, in order to make the assumptions themselves the subject of investigation in a further philosophical chapter.

## 2. Analysis of Special Problems in Probability 15

### 2.1 The Probability Machine

We want to construct an ideal apparatus in such a way that, on the basis of general common sense considerations, anyone would apply the laws of probability distributions to it. We will then attempt to analyze, always relying on natural insight, which aspects lead us to assume such a distribution. Even though we will initially only analyze the conditions of this particular case, we will reach general results; since we will be led to the essential foundations of the problem of probability, we will realize that the example has been no more than an illustration and a helpful guide towards general laws. It will function more or less in the way of a drawing in a geometric proof. Imagine the following apparatus:

A loop of paper tape is run horizontally across two fixed rollers. The tape is lined in the direction of the rollers with narrow, equally wide, black and white stripes and is moved over the rollers with a uniform velocity. A vertical cylinder is fixed at some distance above the tape with a downward opening; it contains a moveable hammer that is pressed downwards with a strong spring. A projectile is fastened loosely in front of the hammer. If the hammer is released
after it has been cocked | and the spring compressed, it will accelerate and shoot 16
the projectile downwards; the projectile will hit the fast moving paper tape and picrcc it. Furthcrmorc assumc that thc apparatus is sct up in a way so that every time a shot is released the right borderline of a white stripe is at that point

des Raumes befindet, wo nachher die Durchbohrung stattfindet, so daß für jeden Schuß die gleichen Anfangsbedingungen bestehen. Würde der Apparat mit vollkommener Präzision arbeiten, so würden stets nur Streifen derselben Farbe getroffen werden; aber die bei der wirklichen Ausführung einer solchen Konstruktion auftretenden Schwankungen verhindern dies. Wenn nun die Geschwindigkeit des Bandes groß genug und die Querstreifen schmal genug sind, so wird der Erfolg der sein, daß nach einer großen Anzahl solcher Schüsse etwa gleichviel schwarze und weiße Querstreifen auf dem Papier durchbohrt sind. Diese Annahme ist ohne weiteres einleuchtend, und wir wollen sie unbewiesen an die Spitze stellen. Es ist zu untersuchen: welches sind die Bedingungen, die wir hierbei stillschweigend vorraussetzen, und wie müssen sie sich ändern, um das Resultat der Gleichverteilung genauer zu machen?

Die erste und einfachste Bedingung ist die, daß der Vorgang des Schießens sehr häufig geschieht, in zwei oder drei Malen wird niemand schon eine angenäherte Gleichverteilung erwarten, dagegen bei hundert Schüssen schon mit größerer Sicherheit, bei tausend mit noch größerer. Unsere Aussage ist überhaupt schon so formuliert, daß sie von einer größeren Anzahl von Schüssen spricht. Nur so kann sie überhaupt eine Aussage vom wirklichen Geschehen sein. Die Behauptung, für einen Schuß sei die Wahrscheinlichkeit, einen weißen Streifen zu treffen, gleich $\frac{1}{2}$, sagt nichts von der Wirklichkeit aus, die kommen wird, sondern läßt sie prinzipiell unbestimmt. Erst bei der Voraussetzung einer größeren Zahl von Schüssen läßt sich die Aussage in die Form kleiden: dies wird geschehen. Auch die Behauptung, daß mit größerer Zahl der Schüsse die Genauigkeit steigt, leuchtet ein; doch wird dieser Gedanke erst später untersucht werden. Es muß hier noch hinzugefügt werden, was mit der Wiederholung „desselben Vorgangs“ gemeint ist. Es ist natürlich nicht richtig, die zahlreichen aufeinander-
17 folgenden Schußvorgänge als identisch | anzusehen, vielmehr ist jeder ein neuer Vorgang. Unter „demselben“ Vorgang sei vielmehr ein solcher verstanden, der sich von den anderen nur durch seine Position in der Zeit unterscheidet, während sämtliche übrigen meßbaren physikalischen Bestimmungsstücke den gleichen Wert haben.

Die zweite und wesentliche Bedingung ist die, daß die beiden Vorgänge, um deren Zusammentreffen es sich handelt – das rollende Band und das fliegende Geschoß – voneinander unabhängig sind. Wenn man sich etwa alle weißen Streifen mit einem Magneten belegt denkt und das Geschoß aus Eisen, so würde es von den weißen Streifen angezogen werden und diese würden häufiger getroffen werden als die schwarzen; derartige Abhängigkeiten müssen ausgeschlossen werden. Nun ist es natürlich richtig, daß es unabhängige Vorgänge in der Natur nicht gibt. Denn es gibt keine abgeschlossenen Systeme, jedes System übt durch seine Oberfläche eine Wirkung auf seine Umgebung aus, sei diese nun thermischer oder mechanischer Natur oder möge sie in der Emission oder Absorption von Strahlung bestellen; es ist prinzipiell unmöglich, die Vorgänge in einem räumlich abgegrenzten System mathematisch darzustellen, ohne es damit gleichzeitig in Verbindung mit sämtlichen Vorgängen des Universums zu bringen. Doch sind die weitaus meisten dieser Einflüsse numerisch verschwindend

where later the piercing will occur,[E3] so that we have the same initial conditions for each shot. If the apparatus worked with perfect precision, then stripes of the same color would be hit every time; but the variations occurring in the actual implementation of such a construction will prevent this from happening. If the speed of the tape is high enough and the stripes on the tape sufficiently narrow, then we will find that after a large number of such shots, about equally many black and white stripes are pierced on the tape. This assumption seems obvious and we will make it at the beginning without proof. We have to investigate: What are the conditions that we tacitly assume here and how must they change to make the resulting distribution more exactly equal?

The first and simplest condition is that the process of shooting has to occur many times. No one will expect an approximately equal distribution after just two or three times, but with higher certainty after one hundred shots and with even higher certainty after a thousand shots. Our claim was already formulated to refer to a large number of shots. Only in this way can it be a claim about an actual event. The claim that for one shot the probability of hitting a white stripe is equal to $\frac{1}{2}$ says nothing about the reality that will follow, but leaves it undetermined in principle. Only on condition of a large number of shots can the claim be expressed in the form: this will occur. The claim that the precision will increase with a large number of shots seems intuitive; but we will analyze this thought later. We must also say what is meant by a repetition of the "same process." Of course it is incorrect to consider the numerous consecutive shooting processes as identical, | rather each one is a new process. Instead, we mean by 17
the "same" process one that only differs from the others in its position in time, while all other measurable physical variables have the same value.

The second and essential condition is that the two events whose co-occurrence is being considered – the moving tape and the flying projectile – are independent. If one imagines all white stripes to be covered by a magnet and the projectile to be made of iron, then it would be attracted to the white stripes and they would be hit more often than the black ones; such dependencies have to be excluded. Of course, it is correct that there are no independent events in nature. There are no closed systems. Every system affects its environment via its surface, be it by mechanical or thermal effects or by the emission or absorption of radiation. It is in principle impossible to represent mathematically the events of a spatially bounded system without at the same time linking it to all other events in the universe. However, the majority of these influences are numerically neg

E3. This is an unusual way to fix a reference point. Reichenbach assumes, as becomes clear later (p. 18 [63]), that the projectile is fired vertically downwards and has no movement in the horizontal direction. Reichenbach then fixes as reference on the tape the right border of a white stripe that passes at the moment of firing exactly vertically below the (center) of the shooting device. Reichenbach appears to assume that the only variation in this device may arise from the projectile's travel time. See p. 21 [67] in the thesis and footnote 15.

gegenüber wenigen anderen. Die Bewegung des rollenden Bandes ist z. B. wesentlich gegeben durch die Geschwindigkeit der Rollen; die gleitende Reibung auf deren Kanten, die leisen Erschütterungen, die ungleiche Erwärmung einzelner Stellen usw., die aus der geradlinigen Bewegung des Bandes in Wahrheit ein Bündel kompliziert gewundener Kurven machen, sind zu vernachlässigen gegenüber der gleichförmigen Translation. So kommt es, daß wir mit gutem Recht die Bewegung des Bandes als Funktion nur weniger Konstanten darstellen; z. B., wenn $x, y, z$ die Koordinaten eines auf dem Band markierten Punktes bedeuten:

$$x = f_x(t, a_1, \ldots, a_n)$$
$$y = f_y(t, a_1, \ldots, a_n)$$
$$z = f_z(t, a_1, \ldots, a_n)$$

Die übrigen Einflüsse, welche neue Konstanten hinzubrächten, oder anders aufgefaßt, diese Konstanten $a_1 \ldots a_n$ zu Funktionen anderer Größen machen wür-
18 den, verschwinden dem gegenüber. | Wir können den Begriff der Abgeschlossenheit und damit der Unabhängigkeit so definieren: Zwei Vorgänge sind unabhängig, wenn ihre Gleichungen[4]

$$x_i = f_i(t, a_1, \ldots, a_n) \quad x'_i = f'_i(t, a'_1, \ldots, a'_m)$$

so beschaffen sind, daß die Größen $x_i$ sich nicht wesentlich ändern, wenn die $x'_i$ andere Werte annehmen, oder daß in den Funktionen

$$a_p = g_p(x'_1, \ldots, x'_i, \ldots) = h_p(t, a'_1, \ldots, a'_m) \quad | \quad p = 1, 2, \ldots, n$$

selbst großen Veränderungen der $x'_i$ nur verschwindend kleine Änderungen der $a_1, \ldots,\ a_n$ entsprechen, und das für alle Werte von $x'_i$.

Es ist leicht zu sehen, daß diese Bedingung von unserem Apparat erfüllt wird. Das Abrollen des Bandes ist ganz unabhängig davon, an welcher Stelle des Raumes sich das Geschoß befindet. Die Gleichungen eines auf dem Band festen Punktes sind:

$$x = ct + a \qquad y = b = 0 \qquad z = d = 0 \tag{2.1}$$

wenn die $x$-Achse des dreiachsigen Koordinatensystems in die Bewegungsrichtung des Bandes gelegt wird; die Vertikalachse möge die $z$-Achse sein. $c$ ist die Geschwindigkeit des Bandes. Die Gleichungen des Geschosses lauten

$$z' = -\frac{1}{2}gt^2 - vt + h \qquad x' = k = 0 \qquad y' = l = 0 \tag{2.2}$$

wo $v$ die Auswurfsgeschwindigkeit des Geschosses, $h$ seine Höhe in diesem Augenblick und $g$ die Erdbeschleunigung ist. Nun ist es klar: die Größen $c$ und $a$ der ersten Gleichung sind unabhängig von $z'$, und umgekehrt ist natürlich auch

4. Die $x_i$ und $x'_i$ bedeuten irgendwelche den Vorgang bestimmenden Größen.

ligible compared with a few others. For example, the movement of the tape is essentially given by the speed of the rollers. The sliding friction on their edges, the slight vibration, the unequal warming of particular parts etc. are factors that in fact make the linear movement of the tape into a bundle of complicated curves, but are negligible compared to the uniform translation. Consequently, we are justified in representing the movement of the tape as a function of just a few constants; e. g., if $x$, $y$, $z$ are the coordinates of a point marked on the tape:

$$\begin{aligned} x &= f_x(t, a_1, \ldots, a_n) \\ y &= f_y(t, a_1, \ldots, a_n) \\ z &= f_z(t, a_1, \ldots, a_n) \end{aligned}$$

The remaining influences that new constants would add, or to put it differently, that would change the constants $a_1, \ldots, a_n$ to functions of other variables, vanish in comparison. | We can define the concept of closedness and consequently 18
of independence as follows: Two processes are independent if their equations[4]

$$x_i = f_i(t, a_1, \ldots, a_n) \quad x'_i = f'_i(t, a'_1, \ldots, a'_m)$$

are such that the variables $x_i$ do not change significantly when the $x'_i$ take different values, or when in the functions

$$a_p = g_p(x'_1, \ldots, x'_i, \ldots) = h_p(t, a'_1, \ldots, a'_m) \quad | \quad p = 1, 2, \ldots, n$$

even large changes of the $x'_i$ only correspond to vanishingly small changes of $a_1, \ldots, a_n$, for all values of $x'_i$.

It is easy to see that our apparatus satisfies this condition. The rolling motion of the tape is independent of the location in space of the projectile. The equations for a fixed point on the tape are:

$$x = ct + a \qquad y = b = 0 \qquad z = d = 0 \tag{3.1}$$

if the $x$-axis of the three axis coordinate system is in the direction of movement of the tape and the vertical axis is the $z$-axis and $c$ is the speed of the tape. The equations of the projectiles are

$$z' = -\frac{1}{2}gt^2 - vt + h \qquad x' = k = 0 \qquad y' = l = 0 \tag{3.2}$$

where $v$ is the exit speed of the projectile, $h$ is its height in that instant and $g$ is the acceleration due to gravity. Now it is clear: the variables $c$ and $a$ in the first equation are independent of $z'$ and conversely, $z'$ is independent of $c$ and $a$.

4. $x_i$ and $x'_i$ stand for any [two] variables determining the process.

$z'$ unabhängig von $c$ und $a$. Das heißt, es ließen sich allerdings Funktionen aufstellen[E1]

$$\begin{aligned} a &= g_1(x', y', z') \qquad & b &= g_2(x', y', z') \\ c &= g_3(x', y', z') \qquad & d &= g_4(x', y', z') \end{aligned} \tag{2.3}$$

weil doch eine Einwirkung tatsächlich stattfindet, es wird etwa durch den dem Geschoß vorangehenden Luftstrom das Band etwas nach unten gedrückt, so daß die $z$-Komponente seiner Bewegung nicht mehr 0 ist – aber selbst bei großen Änderungen von $z'$ ändern sich $x$, $y$, $z$ so verschwindend wenig, daß man von Unabhängigkeit reden kann.

19 Die dritte Bedingung ist offenbar die, daß eine gemeinsame Wirkung der beiden unabhängigen Vorgänge zustande kommt. Wenn das Geschoß an einer ganz anderen Stelle des Raumes flöge als dort, wo das Band sich bewegt, so würde überhaupt gar kein Ereignis vorliegen, das wir der Wahrscheinlichkeitsrechnung unterwerfen können. Ja, wenn das Geschoß auch nur dicht neben dem Band vorbeifliegen würde, so wäre nichts da, was wir abzählen könnten; erst das Durchbohren des Bandes schafft eine Wirkung, ein Geschehnis, dessen Häufigkeit wir der Wahrscheinlichkeitsbetrachtung unterziehen können. Man könnte einwenden, auch in diesem Falle ließe sich noch eine Wahrscheinlichkeitsbeobachtung anstellen, indem man etwa mit einem Fernrohr von der Seite visierte und feststellte, was für einen Querstreifen das Geschoß im Augenblick seines Vorüberfliegens bedeckt. Das gäbe allerdings auch ein Wahrscheinlichkeitsexperiment. Nur sind hier noch Zwischenglieder in die Kette der Vorgänge eingeschoben, und deren Zusammentreffen ist dann das schließlich beobachtete Ereignis. Der Lichtstrahl, der von dem Querstreifen durch das Fernrohr in das Auge des Beobachters gesandt wird, wird im allgemeinen unabhängig von der Bewegung des Geschosses seinen Weg nehmen; nur in dem Augenblick, wo das Geschoß vorüberfliegt, trifft er auf ein undurchlässiges Medium und wird vernichtet resp. reflektiert. Hier sind der Lichtstrahl und das fliegende Geschoß die beiden Vorgänge, auf deren unmittelbares Zusammentreffen es ankommt. In allen Fällen, wo es sich scheinbar gar nicht um solche gemeinsame Wirkung handelt, werden sich doch die betreffenden Zwischenglieder leicht aufzeigen lassen.

Die gemeinsame Wirkung besteht aber in nichts anderem, als daß für einen Augenblick die beiden Vorgänge voneinander abhängig werden. Während der sehr kleinen Zeit $\Delta t$, während der das Papier durchbohrt wird, werden die Gleichungen 2.1 und 2.2 nicht mehr gelten, sondern die Gleichungen 2.3 werden jetzt auf einmal wesentliche Änderungen der Konstanten herbeiführen. Das Papier wird in einem kleinen Gebiet von dem Geschoß nach unten ausgebaucht werden; die $z$-Komponente seiner Bewegung, die vorher 0 war, wird also eine meßbare Funktion von $z'$. Ebenso wird das Geschoß von dem Papierband ein wenig nach der Seite mitgerissen werden, d. h. seine Komponente $x'$ wird eine Funktion von $c$. Da die Einwirkung des Bandes auf das Geschoß durch moleku-
20 lare | Kräfte zustande kommt, wird während dieser Zeit die Beschleunigung des

E1. In the German original there is an error in the equation for $c$. It says: $c = g_3(z', y', z')$.

However, it is possible to construct functions

$$a = g_1(x', y', z') \qquad b = g_2(x', y', z') \qquad (3.3)$$
$$c = g_3(x', y', z') \qquad d = g_4(x', y', z')$$

since there is in fact an effect, for example the tape is pressed downwards slightly by the air pressure in front of the projectile, so that the $z$-component of its movement is not 0 anymore – but even in the case of large changes of $z'$, the changes in $x, y, z$ are vanishingly small, so that one can still speak of independence.

Evidently, the third condition is that the two independent processes result 19
in a combined effect. If the projectile traveled in a completely different point in space than that where the tape is moving, no event would have taken place that we could consider in a probability calculus. Even if the projectile just missed the tape, there would be nothing we could count; only the piercing of the tape creates an effect, an event, whose frequency we can analyze with probabilities. One may argue that even in this case a probability observation could still be obtained by attaching a telescope sideways and determining which stripe the projectile covers in the instance it passes by the tape. This would also constitute a probability experiment, only that here intermediary links are inserted into the chain of events and their co-occurrence is then the observed event. The ray of light that travels from the stripe through the telescope to the eye of the observer will generally travel independently of the path of the projectile; only in the instant when the projectile passes by does it hit an impermeable substance and is destroyed or reflected. In this case, the ray of light and the flying projectile are the processes on whose co-occurrence the experiment depends. In all cases where there does not appear to be such a joint effect, it is nevertheless easy to identify the relevant intermediary links.

However, the joint effect consists of nothing more than the fact that for an instant the two processes become dependent on each other. In the short time interval $\Delta t$, in which the paper tape is pierced, the equations 3.1 and 3.2 will not hold any longer, rather the equations 3.3 will now bring about substantial changes in the constants. The paper will be bent downwards in a small area by the projectile; the $z$-component of its movement that used to be 0 will be a measurable function of $z'$. Similarly, the paper tape will drag the projectile sideways slightly, i. e., the projectile's $x'$-component becomes a function of $c$. Since the ef-
fect of the tape on the projectile is caused by molecular forces, | the acceleration 20

Geschosses nicht mehr $g$ sein, sondern durch die Resultierende aus den Molekularkräften und der Erdkraft bestimmt werden. All dies dauert nur eine sehr kurze Zeit, aber doch von endlicher Größe; nachher sind dann die Bewegungen wieder unabhängig voneinander. Dies letztere ist übrigens gleichgültig; was nach dem Eintreffen der Wirkung geschieht, interessiert nicht mehr, da die Entscheidung bereits getroffen ist. Man ersieht daraus, daß man das Zeitteilchen $\Delta t$ beliebig klein wählen kann, wenn man nur die beim ersten unmittelbaren Zusammentreffen entstehende Wirkung wahrnehmen und zählen kann.

Eine vierte Bedingung ist nun hinzuzufügen. Es ist das, was in der Teilung des Bandes in weiße und schwarze Streifen zum Ausdruck kommt. Ganz offenbar wäre von Wahrscheinlichkeitsverteilung der Löcher auf dem Band nicht zu reden, wenn diese sich nicht in Klassen einteilen ließen; hier sind zwei Klassen gewählt; es könnten ebensogut drei oder mehr durch verschiedenfarbige Streifen dargestellte Klassen sein. Wir stehen hier vor dem Punkte des Problems, der als die Frage der gleich möglichen Fälle eine so wichtige Rolle spielt und der auch Kries zu seiner Theorie der Spielräume geführt hat. Man erkennt, daß für unseren ja absichtlich sehr einfach gewählten Fall nicht nur die Frage der Gleichmöglichkeit sehr leicht zu entscheiden ist – offenbar brauchen die schwarzen und weißen Streifen nur gleich breit zu sein, dann ist alles erfüllt –, sondern daß sie auch für die ganze Untersuchung verhältnismäßig gleichgültig ist. Wichtige physikalische Eigenschaften besaßen das fallende Geschoß, das rollende Band; über ihre physikalische Natur mußten Voraussetzungen getroffen werden. Die Teilung in schwarze und weiße Streifen dagegen bedeutet nicht mehr als ein Schema, nach dem man die eingetroffenen Ereignisse gruppiert, gar keine körperliche Eigenschaft der Dinge; welche Farbe die Streifen haben, ist gewiß ganz nebensächlich, ebenso wie ihre absolute Breite, wenn nur schwarze und weiße Streifen stets gleich breit sind – sie liefern uns nicht mehr als ein Zählverfahren für die Löcher auf dem Papier; und wenn nachher die Aussage gefällt wird, daß ebensoviel weiße wie schwarze Streifen getroffen sind, so ist das weniger eine Aussage über die Streifen, als vielmehr über die Natur des Schußvorgangs. Wür-
21 de der Apparat mit absoluter Präzision arbeiten, würde ins|besondere das Geschoß stets die gleiche Zeit brauchen, bis es das Band berührt, so würde offenbar stets ein Streifen der gleichen Farbe getroffen werden; denn es war ja festgesetzt, daß zu Beginn eines Schusses gerade der rechte Grenzstrich eines weißen Streifens die Stelle passieren soll, wo nachher die Durchbohrung stattfindet. Erst dadurch, daß das Geschoß jedesmal eine etwas andere Fallzeit braucht, kommt eine Wahrscheinlichkeitsverteilung zustande. Über den Schußvorgang wird etwas behauptet, es wird festgestellt, daß für die Schwankungen seiner Dauer ein ganz bestimmtes Gesetz gilt, welches seinen anschaulichen Ausdruck in der Tatsache findet, daß gleichviel weiße und schwarze Streifen getroffen werden. Mehr als eine einleuchtende Illustration für irgendein Naturgesetz in dem zeitlichen Ablauf des Schusses ist die Gleichverteilung nicht. Es wird sich deshalb darum handeln, eine exakte Formulierung dieses Gesetzes zu finden.

Der Einfachheit halber denken wir uns vor der Zählung der Löcher noch eine Verschiebung vorgenommen. Die Anfangspunkte, die für jeden Schuß auf

of the projectile will not be $g$ at this stage anymore, but will be determined by the resulting force given by the acceleration due to gravity and the molecular forces. All this takes a very short, but finite, time; afterwards the movements are independent of one another again. This last point is irrelevant; whatever takes place after the occurrence of the effect is of no interest, since the decision has already been made. It follows that the time interval $\Delta t$ can be chosen arbitrarily small, as long as one can observe and count the effect of the first direct co-occurrence.

A fourth condition has to be added: It is what is expressed by the fact that the tape is divided into black and white stripes. Evidently it is impossible to speak of a probability distribution over the holes in the tape if one cannot group them into classes; in this case we chose two classes, but there may as well be three or more classes represented by different colored stripes. Here we are confronted with the part of the problem that plays an important role with respect to equi-possible events and that led Kries to his theory of event spaces. Note that in our case, which was deliberately chosen to be simple, the question of equi-possibility is not only easy to decide – evidently the black and white stripes only need to be equally wide, then everything is satisfied – but the question is also relatively insignificant to the whole investigation. The falling projectile and the moving tape have important physical properties; we had to agree on conditions regarding their nature. However, the division into black and white stripes is no more than a schema according to which events are grouped, not a physical property of the things; obviously the color of the stripes is irrelevant, as is their absolute width, as long as black and white stripes always have the same width – they provide no more than a counting procedure for the holes in the paper. If in the end the claim is made that an equal number of black and white stripes were hit, then this is not a claim about the stripes, but rather about the nature of the shooting process. If the apparatus worked with perfect precision, | in particular, if the projectile al- 21
ways required the same amount of time until it touched the tape, then a stripe of the same color would be hit each time; after all, we assumed that at the beginning of a shot, the right border line of a white stripe should pass the point where later the piercing will take place. A probability distribution only appears because the projectile requires a slightly different drop time on each occurrence. A claim is made about the shooting process, we detect that the variation of the drop time is subject to a particular law, which becomes visually apparent in the fact that an equal number of black and white stripes are hit. The equal distribution is no more than a plausible illustration of a law of nature present in the temporal sequence of events of the shot. Consequently we will be concerned with a precise formulation of this law.

For simplicity we will assume a further mapping prior to the counting of the holes. The starting points that are assumed for every shot to be fixed to a point of

einen bestimmten Punkt von relativ gleicher Lage festgelegt waren, sollen derart verschoben werden, daß sie alle in einem einzigen solchen Anfangspunkt zusammenfallen. Dabei wird nichts Wesentliches geändert; die Löcher werden entsprechend verschoben, behalten aber alle ihre Lage relativ zu den Querstreifen. Insbesondere bleiben sie alle auf Streifen der gleichen Farbe wie vorher. Dabei werden alle Löcher in einem Gebiet des Bandes konzentriert; solche, die früher weit auseinander lagen, liegen jetzt vielleicht auf demselben Querstreifen. Man erkennt jetzt, daß es eine hinreichende Bedingung für die Gleichverteilung ist, wenn auf jedem Streifen annähernd ebensoviel Löcher liegen wie auf dem ihm unmittelbar benachbarten. Mögen dann auch weit entfernte Streifen eine ganz verschiedene Anzahl von Löchern aufweisen, die Anzahl der auf sämtlichen schwarzen Streifen gelegenen Löcher wird nahezu gleich sein der Zahl der auf sämtlichen weißen gelegenen.

Diese Annahme bedeutet aber nichts anderes, als daß alle Werte der Schußdauer mit annähernd der gleichen Häufigkeit vorkommen wie benachbarte Werte. Denn die Entfernung der Löcher vom Anfangspunkt ist ein direktes Maß für die Länge der Schußzeit. Um unsere Aussage über die Häufigkeit der Zeitwerte
22 | zu präzisieren, wählen wir jetzt die dafür in der Wahrscheinlichkeitsrechnung allgemein übliche Form.

Wir definieren eine Funktion

$$y = \phi(x)$$

derart, daß die Anzahl der Werte $x$ für die Schußzeit, die in ein beliebiges Intervall $a$ bis $b$ fallen, gleich ist

$$N \int_a^b \phi(x)\, dx$$

wo $N$ die Gesamtzahl der überhaupt vorkommenden Schußzeiten bedeutet. Es läßt sich dann beweisen,[5] daß die vorher genannte Annahme dann erfüllt ist, wenn $\phi(x)$ eine im Riemannschen Sinne integrierbare Funktion ist. Die Genauigkeit, mit der die Gesamtzahl der auf schwarzen Streifen gelegenen Löcher an die der auf weißen gelegenen heranrückt, läßt sich dann unter eine beliebig vorgegebene Größe bringen, wenn ich die Streifen klein genug wähle. Graphisch läßt sich dies gut versinnlichen: ich trage auf einer Abzissenachse die Werte $x$ für die Schußzeiten nach rechts auf und lasse eine enge Teilung von der Teilbreite $\Delta x$ durch alle hindurchlaufen. Über jedem Punkt sei der zugehörige Wert $\phi(x)$ als Ordinate aufgetragen, dann wird der einem Rechteck ähnliche Streifen

$$\int_x^{x+\Delta x} \phi(x)\, dx$$

5. Poincaré gibt (*Probabilité*, p. 149) einen Beweis dieses Satzes, der Stetigkeit und Differenzierbarkeit von $\phi(x)$ vorraussetzt. Dies ist jedoch nicht notwendig, es ist leicht einzusehen, daß es eine hinreichende Bedingung ist, wenn das Integral eine stetige Funktion seiner oberen Grenze ist. Dies ist aber bereits für die Riemannsche Integrierbarkeit erfüllt. – Poincaré ist meines Wissens der erste, der die Wahrscheinlichkeitsprobleme der Glückspiele auf die Existenz einer derartigen Funktion zurückführte. Vgl. auch Wissenschaft und Methode, 1. Buch, 4. Kapitel.

relatively equal position are to be mapped to coincide in one such starting point. This changes nothing essential; the holes are mapped correspondingly, but retain their position relative to the stripes. In particular they will all remain on a stripe of the same color as they were initially. In this way all holes are concentrated in one part of the tape; those that were far apart initially may now lie on the same stripe. We realize now that it is a sufficient condition for the equal distribution that every stripe contains approximately the same number of holes as the one immediately next to it. Even if separated stripes have a different number of holes, the sum of all holes on black stripes will be approximately equal to the sum of all holes on white stripes.

However, this assumption means nothing more than that all values of the length of the shot-time occur with approximately the same frequency as neighboring values, since the distance of the holes to the starting point is a direct measurement of the length of time of the shot. In order to make our claim about the frequency of the time values | more precise, we now choose the notation com- 22
monly used in probability theory.

We define a function

$$y = \phi(x)$$

such that the number of $x$-values for the time of the shot that fall into an arbitrary interval $a$ to $b$ is equal to

$$N \int_a^b \phi(x)\, dx$$

where $N$ is the total number of time measurements of the shots that occur. It can then be proven[5] that the previously mentioned assumption is satisfied when $\phi(x)$ is a Riemann integrable function. The accuracy by which the total number of holes on black stripes approaches the number of those on white stripes can be bounded by an arbitrarily given quantity if one chooses the stripes small enough. This can be easily illustrated graphically: I denote, starting from the left, the $x$-values for the time measurement of the shot on the $x$-axis and overlay them with a narrow partition of width $\Delta x$. Above each point we enter the corresponding $y$-value $\phi(x)$, then the bar that closely resembles a rectangle,

$$\int_x^{x+\Delta x} \phi(x)\, dx$$

5. Poincaré provides a proof of this claim (*Probabilité*, p. 149) that assumes continuity and differentiability of $\phi(x)$. However, this is not necessary, since it can easily be seen that it is a sufficient condition if the integral is a continuous function of its upper limit. This is already satisfied by the Riemann integration. As far as I know, Poincaré is the first to trace back the problems of probability in games of chance to the existence of such a function. See also *Wissenschaft und Methode* [*Science and Method*], book 1, chapter 4.

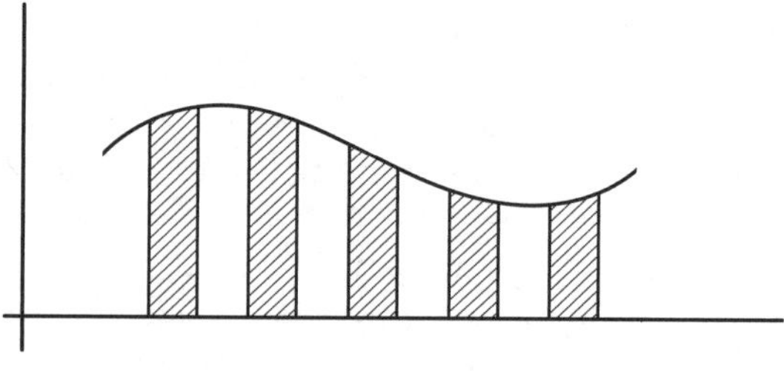

Abbildung 1.

23 bis auf den konstanten Faktor $N$ die Anzahl der Werte $x$ repräsen|tieren, die in das Intervall $\Delta x$ fallen. Die Teilung $\Delta x$ entspricht der Einteilung der Schußzeiten, die durch die Querstreifen des Bandes gegeben wird. Die Anzahl der auf schwarzen Querstreifen gelegenen Löcher ist durch die von den schraffierten Ordinatenstreifen bedeckte Fläche dargestellt, die Anzahl der auf weißen Querstreifen gelegenen durch die von den weißen Ordinatenstreifen bedeckte Fläche. Die Größen dieser Flächen werden einander um so näher kommen, je enger ich die Teilung $\Delta x$ wähle. Dies ist unabhängig vom Anfangspunkt; ich darf die Teilung als Ganzes beliebig nach rechts oder links verschieben, so daß die Endpunkte der Teilchen sich nicht mehr mit den früheren decken, und doch bleibt für alle diese Lagen die Differenz des schraffierten Flächeninhalts und des nicht schraffierten unterhalb derselben kleinen Größe. Nun wird allerdings, da $N$ endlich ist, die wirkliche Verteilung niemals exakt durch die Kurve dargestellt. Dem Intervall $\Delta x$ ist, wenn überhaupt in jedem noch mindestens ein wirklich beobachteter Wert liegen soll, eine untere Grenze gesetzt. Es möge nun $\Delta x$ so gewählt werden, daß in jedem noch eine größere Anzahl Werte $x$ liegt – sie sei gleich h – und über ihm ein Rechteck konstruiert werden, dessen Inhalt gleich $\frac{h}{N}$ ist; dieses Rechteck wird nicht genau dem Ordinatenstreifen gleich sein. Die Einführung der Verteilungsfunktion bekommt nun erst dadurch ihren Sinn, daß man eine Approximation der wirklich beobachteten Verteilung an die durch die Kurve geforderte behauptet; d. h. es muß mit wachsendem $N$ die Teilung $\Delta x$ kleiner gewählt werden dürfen und die Differenz zwischen dem Rechteck und dem Ordinatenstreifen kleiner werden. Der durch die endliche Anzahl von Wiederholungen mit Hilfe der Rechtecke definierte Treppenweg muß sich mit wachsendem $N$ der Kurve $\phi(x)$ anschmiegen.

Es ist jetzt noch eine Bedingung für die Funktion hinzuzufügen. Wenn überhaupt durch eine endliche Zahl von Wiederholungen eine Annäherung an die Kurve zustande kommen soll, so dürfen nicht beliebig viele Intervalle eine beliebig große Häufigkeit aufweisen. Denn dann würde zwar in einer endlichen Zahl von Intervallen die vorgeschriebene Häufigkeit annähernd erreicht werden können, in der unendlichen Menge der anderen Intervalle aber würde die tatsächliche Häufiglteit 0 sein und von einer Konvergenz gegen die geforderte
24 Häufigkeit nicht gesprochen werden | können. Es hätte also gar keinen Sinn, von

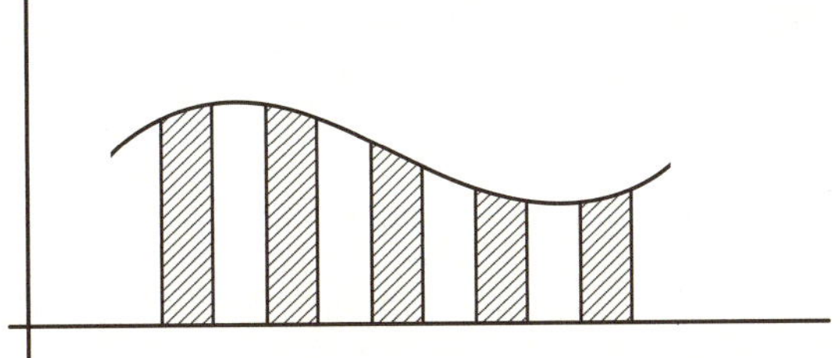

Figure 1.

represents up to the constant factor $N$ the number of the $x$-values that | fall into 23
the interval $\Delta x$. The partition $\Delta x$ corresponds to the classification of the time measurements for the shots that is given by the stripes on the tape. The number of holes on black stripes is represented by the shaded area, the number of holes on white stripes is represented by the unshaded area. The sizes of these areas will converge towards each other the narrower I choose the partition $\Delta x$. This is independent of the starting point; I may shift every partition as a whole arbitrarily to the left or the right, in such a way that the end points of the parts do not coincide with the previous ones and nevertheless for all these shifts the difference of the shaded and unshaded areas will remain bounded by the small given quantity. However, since $N$ is finite, the real distribution will never be represented exactly by the curve. The interval $\Delta x$ is bounded below if every interval is supposed to contain at least one observed value. Let $\Delta x$ be chosen in such a way that every interval contains a larger number of observed $x$-values – say $h$ – and construct a rectangle of area $\frac{h}{N}$ over this interval; this rectangle will not be exactly equal to the ordinate bands of the distribution.[E4] The introduction of the distribution function only has meaning by claiming that the curve approximates the true observed distribution; i. e., as $N$ increases and the partition $\Delta x$ is made smaller the difference between the rectangle and the ordinate band must decrease. The step function defined by the finite number of repetitions using rectangles must hug the curve $\phi(x)$ as $N$ increases.

One further condition for the function has to be added. If any approximation of the curve is supposed to be obtained in a finite number of repetitions, then it cannot be the case that an arbitrary number of intervals possess an arbitrarily high frequency, since then the required frequency could be approximately obtained for a finite number of intervals, but in the infinite set of remaining intervals the actual frequency would be 0 and one would not be able to speak of a convergence towards the required frequency. | It therefore would not make any 24

E4. That is, the number of observed points between $x_i$ and $x_i + \Delta x$ will not equal

$$\int_{x_i}^{x_i+\Delta x} f(x)\,dx.$$

Annählerung zu reden, wenn diese in einer unendlichen Menge von endlichen Intervallen nicht stattfindet. Deshalb muß die Kurve entweder an beiden Seiten begrenzt sein oder asymptotisch gegen die $x$-Achse konvergieren. D. h. die Kurve muß mit der $x$-Achse ein endliches Flächenstück abgrenzen. Mathematisch kommt dies in der Forderung zum Ausdruck, daß das

$$\int_{-\infty}^{+\infty} \phi(x)\, dx$$

einen endlichen Wert haben soll; es bedeutet offenbar die Wahrscheinlichkeit dafür, daß der Wert der Schußzeit überhaupt in dem Intervall $-\infty$ bis $+\infty$ liegt. Dies gilt aber mit Gewißheit, und damit mit wachsendem Intervall überhaupt eine Annäherung an diesen Wert stattfinden kann, muß das Integral einen endlichen Wert haben. Es wird allgemein durch die Wahl der Konstanten in der Funktion gleich 1 gesetzt, entsprechend der Terminologie der Wahrscheinlichkeitsrechnung.

Wir wollen eine solche Funktion, die im Riemannschen Sinne integrierbar ist und mit der $x$-Achse ein endliches Flächenstück abgrenzt, wenn sie einer Reihe von $N$ Wiederholungen derart zugeorgnet ist, daß das Verhältnis der Anzahl $h$ der in das Intervall $a$ bis $b$ fallenden Werte zur Gesamtzahl $N$, also der Bruch $\frac{h}{N}$, mit wachsendem $N$ nach dem Ausdruck

$$\int_{a}^{b} \phi(x)\, dx$$

strebt, eine *Wahrscheinlichkeitsfunktion* nennen. Damit ist als Bedingung der Wahrscheinlichkeitsverteilung die Existenz einer Wahrscheinlichkeitsfunktion aufgezeigt worden. Sie bedeutet lediglich eine Vorraussetzung über die Natur des Schußvorganges und hat mit den übrigen Bestandteilen der Machine nichts zu tun. Sie ist auch ganz unabhängig von der zweiten und dritten Bedingung, denn diese, die Unabhängigkeit der Vorgänge und ihre schließliche gemeinsame Wirkung, könnten erfüllt sein, ohne das eine Variation der Schußzeiten überhaupt vorhanden ist. Die Bedeutung dieser beiden Bedingungen ist eine ganz andere: sie müssen erfüllt sein, wenn die in der vierten Bedingung behauptete
25 Gesetzlichkeit der Variation der Schußzeiten | *deutlich sichtbar* sein soll, wenn sie eine ganz bestimmte physikalische Wirkung besitzen soll. Auch wenn von dem ganzen Apparat nur die Schußvorrichtung vorhanden wäre und in Tätigkeit gesetzt würde, so würde von den Schußzeiten die behauptete Struktur gelten, nur würde sie nicht anschaulich zum Ausdruck kommen. Der gleichen Sichtbarmachung einer Gesetzmäßigkeit dient auch die gleiche Breite der schwarzen und weißen Streifen. Dieser Teil des Apparates hat eigentlich nur die Funktion einer Uhr; er wäre vollkommen dadurch zu ersetzen, daß man mit Hilfe einer Uhr die Länge der Schußzeiten mißt und dann die gleiche Gesetzmäßigkeit feststellt. Und die physikalische Konstruktion der Uhr unterscheidet sich ja auch nicht nennenswert von der hier gegebenen Anordnung. Die Messung der Zeit, die dort durch ein gleichförmig sich drehendes Rad geschieht, wird hier durch

sense to speak of an approximation if it does not occur in an infinite set of finite intervals. This is the reason why the curve has to be bounded from both sides or it has to converge towards the $x$-axis asymptotically, i. e., the curve and the $x$-axis must bound a finite area. Mathematically this can be expressed by the condition that

$$\int_{-\infty}^{+\infty} \phi(x)\, dx$$

must have a finite value; clearly this quantity is the probability that the value of the time of the shot is in the interval $-\infty$ to $+\infty$ at all. However, this is guaranteed, and in order to make an approximation to this value with an increasing interval possible, the integral must have a finite value. It is generally set to be 1 by adjusting the constant in the function just as in the terminology of probability calculus.

We will call a function $\phi$ a *probability function* if it can be integrated by the Riemann method, if it bounds a finite area with the $x$-axis and if it corresponds to a series of $N$ repetitions in such a way that the ratio of the number $h$ of values that fall in the interval $[a, b]$ to the total number $N$, i. e., the fraction $\frac{h}{N}$, approaches the value

$$\int_{a}^{b} \phi(x)\, dx$$

as $N$ increases. We have thereby shown that the condition for a probability distribution is the existence of a probability function. It simply is an assumption about the nature of the shooting process and has nothing to do with the other components of the machine. It is also completely independent of the second and third conditions since these, the independence of the processes and their joint effect, can be satisfied without the presence of a variation in the shooting times. The meaning of these two conditions is completely different: they have to be satisfied if the regularity of the variation of the shooting times given in the fourth
condition is supposed to be | *clearly visible*, if it is supposed to have a particu- 25
lar physical effect. Even if the apparatus only consisted of the firing mechanism, then – when used – the described structure of the shooting times would be present, only it would not become visually apparent. The equal width of the black and white stripes contributes in a similar way to the visualization of the regularity. This part of the apparatus basically only has the function of a clock. If it were entirely replaced by measuring the times of the shots with a clock, we would still find the regularity. After all, the physical constitution of the clock does not differ significantly from the set-up presented here. The time measurement that in the case of the clock is done by a uniformly turning wheel is achieved

das gleichmäßig abrollende Band bewerkstelligt. Die Unabhängigkeit des Uhrwerks ist ebenso erforderlich. Und die Gruppierung der Schußzeiten, die innerhalb eines kleinen Intervalls beieinanderliegende Werte zu einer Schar zusammenzieht, während die in das Nachbarintervall fallenden Werte zur anderen Schar gerechnet werden, wird hier durch die Streifen gegeben, während sie bei Uhrmessungen nachträglich vom Rechner ausgeführt werden müßte. Nur den Vorzug größerer Anschaulichkeit besitzt diese Maschinerie vor der Uhr.

Wenn wir hier also die Bedingungen der Anwendbarkeit von Wahrscheinlichkeitssätzen aufsuchen, so muß zwischen zwei Arten von ihnen unterschieden werden. Erstens muß aufgezeigt werden, welche bestimmte Gesetzmäßigkeit eines Naturvorgangs zugrunde liegen muß, wenn die Resultate der Wahrscheinlichkeitsberechnung richtig sein sollen. Zweitens müssen die anderen Bedingungen festgestellt werden, unter denen diese Gesetzlichkeit überhaupt erst wahrnehmbar wird, indem sie eine physikalische Wirkung hervorbringt. Als solche Bedingungen hatten wir ermittelt: die Unabhängigkeit der beiden Vorgänge, ihre gemeinsame Wirkung (d. h. schließliche Abhängigkeit) und die Aufstellung eines Zählschemas (die Querstreifen), das die in dem Naturvorgang liegende Gesetzmäßigkeit durch Auszählung feststellen läßt. Daß diese Bedingungen der zweiten Art sämtlich empirisch feststellbar sind, erkennt man sofort; die Unabhängigkeit zweier Vorgänge läßt sich durch Erfahrungsgesetze feststellen, und
26 ob ein Zahlschema | den geforderten mathematischen Ansprüchen genügt, läßt sich ebenfalls nur durch Messung erkennen. Die Frage, ob diese – für die Anwendbarkeit der Wahrscheinlichkeitsrechnung ebenfalls notwendigen – Bedingungen erfüllt sind, ist demnach lediglich durch die *Erfahrung* zu entscheiden, aber auch mit der allgemeinen Sicherheit von Erfahrungssätzen überhaupt lösbar. Ganz anders ist das Problem der Bedingungen der ersten Art, die sich hier auf eine einzige Bedingung reduziert haben. Hier muß zuvor die Frage aufgeworfen werden: Ist es überhaupt prinzipiell erfahrungsmäßig erkennbar, daß eine Wahrscheinlichkeitsfunktion existiert? Es scheint ja zunächst möglich, daß wir im Sinne wie wir etwa das Fallgesetz empirisch als ein Naturgesetz aufstellen, auch zu dem Erfahrungssatz gelangen könnten, daß eine Wahrscheinlichkeitsfunktion bestünde. Dann wären sämtliche Behauptungen, die wir auf Grund von Wahrscheinlichkeitsüberlegungen über das Naturgeschehen aufstellen, reine Erfahrungssätze. Dies ist aber gerade das Problem; und es wird später gezeigt werden, daß die letztere Auffassung falsch ist, daß es sich hier vielmehr um ein metaphysisches Prinzip der Naturerkenntnis überhaupt handelt.

## 2.2 Die Glücksspiele

Es ist bisher an einem Beispiel die Theorie der Wahrscheinlichkeitsrechung entwickelt worden; es war auch dort schon bemerkt, daß das besondere Beispiel nur eine Rolle der Illustration übernimmt, und daß es allgemeine Prinzipien sind, zu denen jene Überlegung führt. Doch bedarf diese Behauptung noch der Bestätigung durch Nachprüfung anderer wichtiger Wahrscheinlichkeitsprobleme. Man kann die Fälle, für die die besprochene Wahrscheinlichkeitsmaschine das ideale

here by the uniformly moving tape. The independence of the clockwork would also be required. The grouping of shooting times into small neighboring intervals is determined by the stripes while in the case of time measurement with a clock it would have to be determined by a calculator retrospectively. The only advantage this apparatus has over the clock is visual.

So, if we want to identify the conditions for the application of probability theorems, then we have to distinguish between two types of conditions. First, if the results of the probability calculation are to be correct, it has to be shown which regularity has to underlie the process in nature. Second, the other conditions must be established that enable the observation of the regularity by producing a physical effect. We established the following such conditions: the independence of the two processes, their common effect (i. e., eventual dependence) and the specification of a counting scheme (the stripes) that allows us to identify by enumeration the regularity present in the natural process. The fact that we can empirically determine all these conditions of the second type is immediately evident; the independence of two processes can be determined by empirical laws, and whether a counting scheme | satisfies the mathematical require- 26
ments can also be recognized by measurement. These conditions are equally necessary for the application of probability calculations. Whether these conditions are satisfied can therefore only be decided by *experience*, but due to the general certainty of principles of experience it is decidable at all. The problem is quite different with respect to conditions of the first type that in this case have been reduced to one single condition. Here we first have to raise the following question: Is it in principle possible to recognize empirically that a probability function exists? Initially it seems possible that, in the same way as we empirically establish that the law of free fall is a law of nature, we could also establish the empirical claim that a probability function exists. In that case all claims that we make about natural processes on the basis of probability considerations would be pure claims of experience. However, this is precisely the problem, and we will later show that this is a misconception; rather we are dealing here with a metaphysical principle of the understanding of nature in general.

## 2.2 Games of Chance

So far we have developed the theory of probability calculation on the basis of an example; we realized that the specific example only functions as an illustration leading to general principles. However, this claim needs to be confirmed by comparing it with other important probability problems. The cases for which

Schema vorstellt, unter dem Namen der Glückspiele zusammenfassen, und diese seien zunächst behandelt.

Für den Fall der geworfenen Münze ist die Einordung sehr leicht zu vollziehen. Hier ist der Stoß, der der Münze beim Werfen erteilt wird, sie gleichzeitig in die Höhe schleudert und in Drehung versetzt, der Vorgang, „derselbe" Vorgang, der häufig wiederholt wird. Was dies Wort „derselbe" bedeuten soll, ist
27 oben schon gesagt worden. Auf die Variation der Fallzeiten | kommt es an; ist der Stoß um ein wenig stärker geworden, so wird die Münze nach etwas späterer Zeit niederfallen und inzwischen noch eine Umdrehung ausgeführt haben, so daß jetzt die andere Seite oben liegt. Die beiden Seiten der Münze übernehmen also die Gruppierung der Fallzeiten in zwei Scharen. Das Resultat wird um so besser sein, je rascher sich die Münze im Verhältnis zur Zeitdauer des Niederfallens dreht. Und die implizite Voraussetzung der Behauptung, daß jede Münzseite nahezu gleich oft daran kommen wird, ist wieder die, daß die Variation der Fallzeiten einem ganz bestimmten Gesetz folgt; dem Gesetz nämlich, daß die Zahl, die jedem kleinen Intervall die Häufigkeit der ihm zugehörigen Werte zuordnet, durch eine Wahrscheinlichkeitsfunktion bestimmt wird. – Für das Roulettespiel, soweit es sich um die Gleichwahrscheinlichkeit von rot und schwarz handelt, ist die Betrachtung genau so durchzuführen.

Der im vorigen angeführte Satz über integrierbare Funktionen läßt sich leicht dahin erweitern, daß die Gleichheit der Rechteckssumme auch gilt, wenn ich drei oder mehr aufeinanderfolgende Rechtecke als die Elemente dreier oder mehrerer verschiedener Scharen betrachte. So kommt es, daß sich auch Probleme mit mehr als zwei „möglichen Fällen" jener Betrachtung unterordnen.

Das Würfelspiel ist der einfachste derartiger Fälle. Hier handelt es sich um eine Gruppierung der Stoßintensitäten in sechs verschiedenen Scharen, die mit Hilfe der sechs Würfelseiten zustande kommt. Die Häufigkeitsfunktion, die hier angesetzt wird, ist genau dieselbe wie bei der Münze; wieder ist sie eine von der Variation der Fallzeiten behauptete Gesetzmäßigkeit. Die andere Vorschrift, die man für dieses Spiel aufstellt, daß der Schwerpunkt des Würfels im Mittelpunkt liegen muß und seine Seiten gleichartig beschaffen sein müssen, bezieht sich nicht auf diese Voraussetzung, sondern auf jene zweite Art von Bedingungen, die oben als für die *Messung* der vorausgesetzten Gesetzmäßigkeit wesentlich charakterisiert wurden. Auch wenn man mit einem einseitig beschwerten Würfel wirft, ist dieselbe Struktur der Variation in den Fallzeiten vorhanden. Nur ist das in dem Fallen der Seiten gegebene Zahlschema nicht mehr der geeignete Ausdruck dafür. Wir müssen, um das einzusehen, den Vorgang des Niederfallens des Würfels etwas genauer betrachten. Während des Fallens rotiert der
28 Würfel, er wird den | Boden im allgemeinen zuerst mit einer Ecke berühren. Bei dem homogenen Würfel wird nun – wenn ich der Einfachheit halber von der kinetischen Energie der Rotation absehe, die an der Problemstellung doch nichts ändert – die Seite zu Boden fallen, die im Moment des Berührens den kleinsten Neigungswinkel gegen den Boden besitzt. Wähle ich die Fallzeit ein wenig größer, so wird der Würfel schon wieder mehrere Umdrehungen ausgeführt haben; während dieser Zwischenzeit hat jede Seite ebenso lange wie die andere einmal

the probability machine described above is an ideal schema can be grouped under the heading of games of chance. We will consider these first.

In the case of flipping a coin the categorization is easy. In this case the blow that throws the coin, flings it upwards and sends it spinning, is the "same" often repeated process. The meaning of the word "same" has been explained above. The variation of the drop times | is crucial; if the blow is slightly harder, then 27
the coin will fall down at a slightly later point in time and will have flipped once more, so that the other face is now up. The two sides of the coin divide the drop times into two sets. The faster the coin rotates relative to the drop time, the better the result. And the implicit condition for the claim that each side of the coin will come up approximately the same number of times is again that the variation of the drop times follows a particular law, namely the following: the number that assigns to each small interval the frequency of the values corresponding to it is determined by a probability function. In the case of roulette, similar considerations apply as far as the equal probability of red and black is concerned.

The theorem about integrable functions mentioned above can easily be extended so that the equality of the sum of rectangles still holds if one considers three or more consecutive rectangles as the elements of three or more different sets. Hence it follows that problems with more than two "possible cases" can be treated in the same way.

The throw of dice is the simplest of such cases. In this case, it is a matter of a grouping of strengths of throws into the six different sets that are due to the six sides of the die. The frequency function that is applied here is exactly the same as with the coin flip; again, it asserts a regularity present in the variation of the throw times. The other condition required for the game, that the center of gravity of the die has to lie at the center of the die and that all its sides are similarly constituted, is not part of this condition. It belongs rather to the second type of condition, which we considered essential to the *measurement* of the regularity assumed above. Even if one throws a biased die, the same structure of variation is present in the drop times, but the counting scheme given by the appearance of a particular side is no longer the appropriate expression for it. In order to understand this we have to take a closer look at the process of the falling die. During the fall the die rotates; | in general it will touch the ground with a corner first. 28
In the case of a homogeneous die – if I ignore for simplicity the kinetic energy of the rotation, which does not change anything about the problem – the side will come down that has the smallest angle to the ground at the moment of impact. If I choose the drop time a little larger, the die will have completed several more rotations; in this time interval each side will have the smallest angle to the ground

oder mehrere Male den kleinsten Neigungswinkel besessen. Die Zwischenzeit ist also in sechs gleiche Teile geteilt worden; jeder dieser Teile wird in einer anderen Schar gezählt. Das sind genau die Bedingungen wie die oben für das Zählschema entwickelten. Liegt jedoch der Schwerpunkt des Würfels einer Seite näher als den anderen, so wird diese auch noch dann niederfallen, wenn sie nicht mehr den kleinsten Neigungswinkel besitzt, da das Lot vom Schwerpunkt zur horizontalen Auffallsebene doch noch durch sie hindurchgeht; sie wird also längere Zeit als die übrigen Gelegenheit zum Niederfallen haben. Hier sind die Teile der Zwischenzeit nicht mehr gleich, und die Bedingungen für das Zählschema sind nicht mehr erfüllt. Man sieht leicht ein, daß dann vielmehr die Häufigkeitszahlen der am Boden liegenden Seiten sich verhalten werden wie die Teile der Zwischenzeiten; ein derartiger Satz ließe sich ganz analog dem anderen für integrierbare Funktionen ableiten.

Man erkennt, daß für alle sogenannten Glücksspiele dieser Art, bei denen ein Vorgang oft wiederholt wird, die gleich möglichen Fälle durch korrespondierende gleiche *Zeit*intervalle gegeben werden.[6] Die Zuordnung geschieht dadurch, daß, wenn eine bestimmte Zeitgröße (die Fallzeit) innerhalb dieses Intervalls liegt, der Fall mit Notwendigkeit eintritt. Von der Variation dieser Zeitgröße wird dann die Voraussetzung gemacht, daß die die Häufigkeit ihrer einzelnen Werte bestimmende Funktion integrierbar ist und mit der $x$-Achse ein endliches Flächenstück abgrenzt; dann folgt mit Gewißheit die gleiche Häufigkeit des Auftretens aller möglichen Fälle.

Im besonderen Fall, der sogar nicht selten verwirklicht ist, kann diese Funktion die Form

29 $$f(x) = const.$$

haben. Dies kommt auf folgende Weise zustande. Man denke sich einen Zeiger über einem Kreis rotierend, etwa wie beim Roulettespiel; durch eine einmalige Stoßkraft werde der Zeiger angetrieben, so daß seine Schwingung allmählich langsamer wird und er schließlich stillsteht. Wenn man den Umdrehungswinkel in Vielfachen von $2\pi$ mißt, so wird die Wahrscheinlichkeit, daß der bis zum Punkt des Anhaltens gemessene Winkel gerade einen bestimmten Wert $\Omega$ annimmt, eine stetige Funktion von $\Omega$ sein, $= \phi(\Omega)$, die an einer Stelle etwa ein Maximum hat und für größere und kleinere Werte allmählich nach 0 geht. Wenn man nun den Kreis in $n$ gleiche Sektoren $\vartheta_1, \ldots, \vartheta_n$ einteilt (vgl. den Kreis in der Figur S. 30), so wird die Wahrscheinlichkeit, daß der Zeiger im Sektor $\vartheta_1$ anhält, gleich sein der Summe der Wahrscheinlichkeiten, daß $\Omega$ zwischen 0 und $\vartheta_1$, $\Omega$ zwischen $2\pi$ und $2\pi + \vartheta_1$, $\Omega$ zwischen $2 \times 2\pi$ und $2 \times 2\pi + \vartheta_1$ usw., und entsprechend für die anderen Sektoren; d. h. ein Sektor entspricht einer Schar von Intervallen, wie sie in der Figur S. 22 gezeichnet sind (dort sind zwei solche Scharen gezeichnet). Wegen der Stetigkeit des Integrals sind diese Summen einander nahezu gleich, und dies um so besser, je kleiner die Schwankung der Funktion $\phi(\Omega)$ ist in den Intervallen 0 bis $2\pi$, $2\pi$ bis $2 \times 2\pi$ usw., die inbezug auf die

6. An Stelle der Zeitgrößen können in anderen Fällen auch andere Größen treten.

once or several times for the same amount of time as any other. The intermediary time is hence split into six equal parts; each part is counted in a different set. These are exactly the same conditions as we developed for the counting scheme above. If the center of gravity of the die lies closer to one side than to the others, then this side will come down even if it does not have the smallest angle to the ground, since the perpendicular from the center of gravity to the horizontal surface still passes through this side; consequently it will have more time than the other sides to come down. In this case the parts of the time interval are not equal anymore, and the conditions for the counting schema are no longer satisfied. Clearly, the frequencies of the sides touching the surface behave in the same manner as the parts of the time interval; such a claim could be derived analogously to the claim about integrable functions.

Note that in all games of chance of this type, in which the process is repeated often, the equi-possible cases are given by corresponding equal *time* intervals.[6] The coordination[E5] is done in such a way that the event occurs necessarily if a particular time measure (the drop time) falls within this interval. One assumes that the function specifying the frequency of particular, varying values of the measure of drop time is integrable and bounds a finite area with the $x$-axis; then we can conclude with certainty that each possible event occurs with equal frequency.

In the particular case, which is not actually that rare, this function can have the form

$$f(x) = const.$$ 29

This happens in the following way. Consider a pointer rotating above a circle – similar to a roulette game. The pointer is struck by a single impulse, so that its rotation speed decreases and eventually stops. If the angle of rotation is measured in multiples of $2\pi$, then the probability that the angle measured until the rotation stops at a particular value $\Omega$ is a continuous function of $\Omega$, $\phi(\Omega)$. It has a single maximum and tends to 0 either side of the maximum. If the circle is split into $n$ equal sections $\vartheta_1, \ldots, \vartheta_n$ (see circle in Figure 2), then the probability that the pointer stops in sector $\vartheta_1$ will be equal to the sum of probabilities that $\Omega$ is between 0 and $\vartheta_1$, $\Omega$ is between $2\pi$ and $2\pi + \vartheta_1$, $\Omega$ is between $2 \times 2\pi$ and $2 \times 2\pi + \vartheta_1$ etc.; and similarly for the other sectors; that is, a sector corresponds to a set of intervals, as they are drawn in Figure 1 (two such sets are shown). Due to the continuity of the integral, the sums [for the respective sectors] are almost equal – more equal the smaller the variation of the function $\phi(\Omega)$ is within each of the intervals 0 to $2\pi$, $2\pi$ to $2 \times 2\pi$, etc., that appear as a period with

6. In other cases the time measures may be replaced with other variables.

E5. We follow here the standard translation of Reichenbach's term "Zuordnung". However, "Zuordnung" may be better understood as "assignment".

Auszählung als Periode auftreten. Wir können nun einerseits diese Schwankung dadurch herabdrücken, daß wir unter Beibehaltung der Sektoren $\vartheta_1, \ldots, \vartheta_n$ die Periode verkleinern und anstatt $2\pi$ kleinere Sektoren als Periode einführen, wie dies z. B. beim Roulettespiel geschieht, wo sich die Periode von einem roten und einem schwarzen Sektor bereits auf dem Kreis vielfach wiederholt. Halten wir aber die Periode $2\pi$ fest, so lassen sich die zu $\vartheta_1, \ldots, \vartheta_n$ gehörenden Wahrscheinlichkeiten dadurch immer weiter annähern, daß wir die Umlaufszeit des Zeigers (oder seine Anfangsgeschwindigkeit) vergrößern; denn damit wird die Wahrscheinlichkeitskurve verflacht und demnach ihre Schwankung in gleichen Intervallen immer kleiner. Wir können also mit beliebiger Genauigkeit erreichen, daß, wenn ich $\vartheta$ von 0 bis $2\pi$ zähle, die Wahrscheinlichkeitsfunktion $f(\vartheta)$ die Form annimmt

$$f(\vartheta) = const.$$

Dabei ist die spezielle Form von $\phi(\Omega)$ gleichgültig.

In diesem Zusammenhang findet sich Gelegenheit, kurz auf die Kriessche Theorie zurückzukommen. Wenn Kries seine Spielräume auf „ursprüngliche“ reduziert, so meint er damit wesentlich nichts anderes als die Zurückführung einer Wahrscheinlichkeitsfunktion von beliebiger Form auf eine solche von der Form $f(x) = const.$ Man denke sich die gleichen Sektoren unseres Kreises durch Verlängerung der Radien auf eine Tangente abgebildet; praktisch läßt sich dies etwa ausführen, indem man in Richtung des rotierenden Zeigers einen Lichtstrahl hinaussendet. (Vergleiche die Figur.) Die Wahrscheinlichkeit, daß

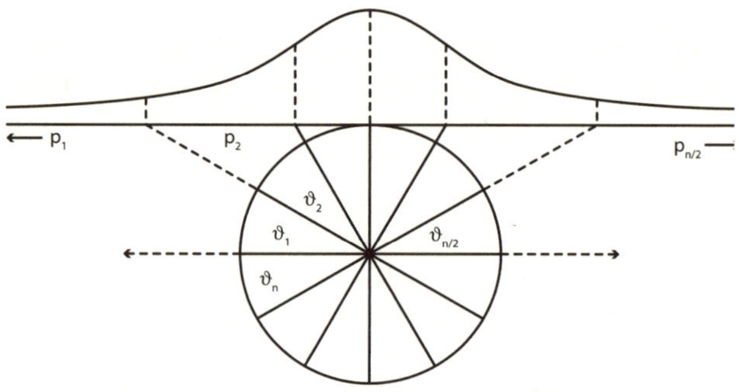


Abbildung 2.

der Lichtpunkt in einem der Intervalle $p_1, p_2, \ldots, p_{n/2}$ anhält, ist für alle diese Intervalle gleich, weil sie für die ihnen korrespondierenden $\vartheta_1, \ldots, \vartheta_{n/2}$ gleich ist; dabei werden aber die $p_1, p_2, \ldots, p_{n/2}$ nach außen immer größer. Die Wahrscheinlichkeitsfunktion $g(p)$ würde also die Bedingung erfüllen müssen, daß die Flächenstücke über den Intervallen $p_1, p_2, \ldots, p_{\frac{n}{2}}$ einander gleich sind; sie würde die Form der über der Tangente als Abszissenachse gezeichneten Kurve haben. Kries' Gedanke ist nun der, daß man deshalb in der Funktion $g(p)$ nicht gleiche Spielräume abteilen dürfe, sondern daß man die den „ursprünglichen“ Spielräumen $\vartheta_1, \ldots, \vartheta_{\frac{n}{2}}$ korrespondierenden Spielräume $p_1, p_2, \ldots, p_{\frac{n}{2}}$

respect to the count. On the one hand we can reduce this variation while maintaining the sectors $\vartheta_1, \ldots, \vartheta_n$ by reducing the period and introducing instead of $2\pi$ smaller sectors as period, e. g., as it is done in the roulette game, where the period of a red and a black sector is repeated several times on the circle. But if we fix the period at $2\pi$, then the probabilities associated with $\vartheta_1, \ldots, \vartheta_n$ can be made to converge by increasing the rotation time of the pointer (or its initial velocity); since this flattens the probability curve and thereby decreases the variation within the equal intervals. Therefore we can ensure with arbitrary precision that when I count $\vartheta$ between 0 and $2\pi$ that the probability function $f(\vartheta)$ takes the form

$$f(\vartheta) = const.$$

The particular form of $\phi(\Omega)$ is irrelevant.

In this context, we can take the opportunity briefly to return to Kries's the- 30
ory. When Kries reduces his events to "elementary" ones, he essentially means nothing other than re-expressing a probability function of an arbitrary form as one of the form $f(x) = const.$ Consider the same sectors of our circle to be projected onto a tangent by extending the radii; in practice this could be achieved by emitting a ray of light in the direction of the pointer. (See figure 2.) The

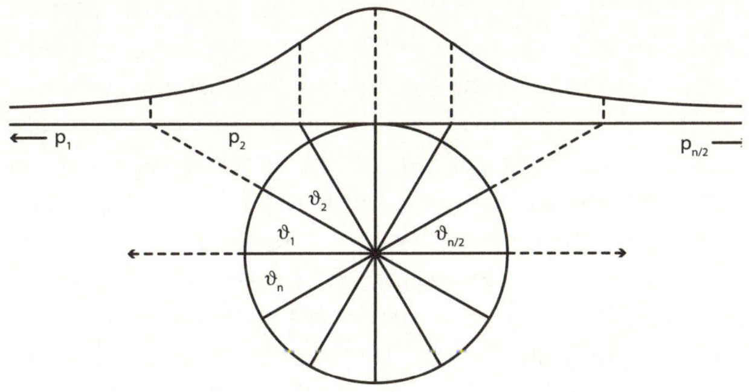


Figure 2.

probability that the light spot stops in one of the intervals $p_1, p_2, \ldots, p_{n/2}$ is the same for all intervals, because it is the same for the corresponding $\vartheta_1, \ldots, \vartheta_{n/2}$; however the $p_1, p_2, \ldots, p_{n/2}$ increase in size outwards. So the probability function $g(p)$ would have to satisfy the condition that the areas above the intervals $p_1, p_2, \ldots, p_{n/2}$ are equal to each other, it would have the shape of the curve drawn as $x$-axis on the tangent. Kries's thought is that one consequently cannot partition the function $g(p)$ into equal events, but rather that one has to consider

als gleichwahrscheinlich ansetzen müsse. Er übersieht nur, daß dies gar nicht nötig ist; wenn man die Funktion $g(p)$ nur in genügend kleine gleiche Intervalle teilt, so läßt sich wegen der Stetigkeit der Abbildung, die $g(p)$ zu einer Funktion mit stetigen Integral gemacht hat, auch $g(p)$ wieder in Scharen von gleicher
31 Wahrscheinlichkeit zerlegen. Auf | die „Ursprünglichkeit“ der Spielräume kann deshalb verzichtet werden.

Auch von „Indifferenz“ der Spielräume haben wir nicht zu reden brauchen; dieses Bestimmungsstück ist gleichfalls durch das stetige Integral überflüssig geworden. Damit ist auch das letzte subjektive Element aus dem Prinzip ausgeschaltet worden; und die mathematische Formulierung hat das Prinzip nicht nur begrifflich verschärft und vereinfacht, sondern auch von fehlerhaften Elementen befreit. Hier liegt die Überlegenheit des Prinzips der Wahrscheinlichkeitsfunktion gegenüber dem Prinzip der Spielräume.

In der alten Darstellung würde man – und dies ist eben der Kriessche Begriff der Indifferenz – die gleiche Wahrscheinlichkeit gleich großer Kreissektoren in dem hier gegebenen Beispiel mit dem Prinzip des mangelnden Grundes begründen. Wegen der Symmetrie, so würde man argumentieren, erscheint kein Sektor vor dem anderen bevorzugt, darum ist die Wahrscheinlichkeit für alle gleich. Das neue Prinzip ermöglicht es uns, auf den Symmetrieschluß zu verzichten, indem es eine Stetigkeitsforderung an seine Stelle setzt. Dies läßt sich ganz allgemein durchführen; stets wenn bei den sogenannten geometrischen Wahrscheinlichkeiten die gleiche Wahrscheinlichkeit zweier Raumteile auftritt, läßt sich – sobald man sich nur die Wahrscheinlichkeitsverteilung durch irgendeine physikalische Konstruktion ausgeführt denkt –, eine Größe angeben, deren Wiederholung in der Zeit oder im Raume durch eine Funktion mit stetigem Integral geregelt ist, und durch Einteilung der Funktion in Scharen von gleichen Intervallen des Arguments folgt dann aus der Stetigkeit die gleiche Wahrscheinlichkeit der Raumteile. Für die geometrische Wahrscheinlichkeit besteht eine Besonderheit nur darin, daß die den Scharen von Intervallen zugeordneten Ereignisse gleichzeitig einer bestimmten, meist sehr einfachen Einteilung des Raumes zugeordnet sind, während dies bei den anderen Problemen wegfällt. Allen gemeinsam gilt, daß die sonst unvermittelt nebeneinanderstehenden gleichwahrscheinlichen Fälle sich einem stetigen Integral in eigentümlicher Weise zuordnen lassen, wodurch es gelingt, die Frage nach der Wahrscheinlichkeit zurückzuführen auf die Frage nach der Existenz dieser Funktion.

## 32 2.3 Das Theorem der zusammengesetzten Wahrscheinlichkeit

Für die Wahrscheinlichkeit, daß zwei voneinander unabhängige Ereignisse gleichzeitig eintreffen, gibt die Wahrscheinlichkeitsrechnung als Regel an, daß sie gleich dem Produkt der Einzelwahrscheinlichkeiten ist. So ist die Wahrscheinlichkeit, mit zwei Würfeln die Seiten 1 und 3 zu werfen, gleich $\frac{1}{6} \times \frac{1}{6} = \frac{1}{36}$. Entsprechend ist, wenn $f_1(x)$ und $f_2(y)$ die Wahrscheinlichkeitsfunktionen zweier verschiedener unabhängiger Ereignisse sind, die Wahrscheinlichkeit einer Kombination bestimmter Werte durch das Produkt $f_1(x)f_2(y)$ gegeben.

as equi-probable the events $p_1, p_2, \ldots, p_{n/2}$, that correspond to the "elementary" events $\vartheta_1, \ldots, \vartheta_n$. What he does not realize is that this is not necessary; if one partitions the function $g(p)$ into small enough equal intervals, then – due to the continuity of the projection that made $g(p)$ to a function with a continuous integral – $g(p)$ can be divided into sets of equal probability. | Consequently one 31
can forgo the requirement that the events be "elementary".

We also do not have to speak of the "indifference" of the events; this aspect has also become superfluous due to the continuous integral. We have thereby managed to eliminate the last subjective component from the principle; the mathematical formulation of the principle has not only made the terminology more precise and simple, but has also removed erroneous elements. This shows the superiority of the principle of the probability function over the principle of event spaces.

In the old representation one would – and this is exactly Kries's concept of indifference – explain the equal probability of equal sized circle sectors in this example with the principle of insufficient reason. Due to the symmetry, one would argue, no sector seems privileged over another, and consequently they all have the same probability. The new principle allows us to forgo the symmetry inference by replacing it with a continuity condition. This can be generalized; whenever in the case of so-called geometric probabilities two spaces have equal probability, then – if one thinks of the probability distribution as generated by some physical process – a quantity can be specified whose reoccurrence in time or space is governed by a function with a continuous integral, and by partitioning the function into sets of equal intervals of the argument value the equal probability of the spaces follows from the continuity. For geometric probability there is only one distinctive feature; the events corresponding to a set of intervals also correspond to a particular, generally very simple, partition of the space, whereas this does not hold for other problems. All have in common that cases which would otherwise be equi-probable by coincidence can be assigned to a continuous integral in a specific way, and thereby the question of probability can be reduced to the question of the existence of this function.

## 2.3 The Composite Probability Theorem 32

Probability calculus specifies the rule that the probability that two independent events co-occur is the product of the probabilities of the two individual events. Consequently, the probability of throwing a 1 and a 3 with two dice is equal to $\frac{1}{6} * \frac{1}{6} = \frac{1}{36}$. Accordingly, if $f_1(x)$ and $f_2(y)$ are the probability functions of two different independent events, then the probability of a combination of particular values is given by the product $f_1(x) f_2(y)$.

Diese Aufstellung läßt sich nicht aus der Existenz der Wahrscheinlichkeitsfunktion ableiten. Es wäre möglich, daß jedes einzelne Ereignis in seiner Wiederholung genau seine Wahrscheinlichkeitsfunktion befolgt, und daß dennoch die Verteilung der Kombinationen keineswegs durch das Produkt geregelt ist. So könnte beim Werfen mit zwei Würfeln jedesmal auf beiden die gleiche Seite oben liegen; dies schließt nicht aus, daß für jeden Würfel jede Seite gleich häufig darankommt, aber die Häufigkeit jeder Kombination wäre dann nicht durch die Zahl $\frac{1}{36}$ gegeben. Die Voraussetzung, daß für jedes einzelne Ereignis eine Wahrscheinlichkeitsfunktion existiert, ist deshalb nicht hinreichend für die Begründung des Gesetzes der zusammengesetzten Wahrscheinlichkeit.

Jedoch läßt sich dieses Gesetz dann begründen, wenn man annimmt, daß auch für die Kombinationen von $x$ und $y$ eine Wahrscheinlichkeitsfunktion $f(x, y)$ existiert, d. h. wenn man annimmt, daß eine Funktion existiert, die jeder Kombination von Werten $x$ aus dem Intervall $a$ bis $b$ und Werten $y$ aus dem Intervall $c$ bis $d$ ihre Häufigkeit zuordnet durch den Ausdruck

$$\int_a^b \int_c^d f(x, y)\, dx\, dy$$

Aus der Integrierbarkeit dieser Funktion läßt sich in ganz ähnlicher Weise,
wie dies früher für die Funktion *einer* Veränderlichen dargestellt wurde, nach-
33 weisen, daß, wenn ich die $X$, | $Y$-Ebene in eine große Zahl kleiner Rechtecke
$\Delta x \Delta y$ einteile und die über diesen errichteten, bis zur Fläche $f(x, y)$ reichen-
den Säulen, die die zugehörige Wahrscheinlichkeit darstellen, in entsprechende
Scharen einteile, die Summen dieser Scharen in der Grenze gleich groß werden.
Die gleich möglichen Fälle entsprechen jeder einer solchen Schar, und aus der
bloßen Existenz einer solchen Wahrscheinlichkeitsfunktion folgt wie früher die
gleiche Häufigkeit dieser Fälle. So ist die Existenz einer solchen Funktion hinreichende Bedingung für die gleiche Häufigkeit aller Kombinationen beim Werfen mit zwei Würfeln.

Die erste Voraussetzung verlangt von dieser Funktion jedoch eine besondere Gestalt; sie fordert nämlich, daß in einer durch dies Gesetz bestimmten Reihe solcher Kombinationen gleichzeitig jede Werteschar $x$ und jede Schar $y$ durch ihre besondere Wahrscheinlichkeitsfunktion geregelt ist. Denken wir uns den Versuch ausgeführt, indem zwei verschiedene Größen $x$ und $y$ zu wiederholten Malen gleichzeitig physikalisch realisiert werden, so werden die verschiedensten Kombinationen $x$, $y$ beobachtet werden. Jede beobachtete Kombination möge durch einen Punkt in einer $x$-$y$-Ebene markiert werden. Wenn man nun die Größe $y$ gewaltsam auf ein Intervall $c$–$d$ beschränkt, so werden diese Punkte alle in einem Streifen parallel zur $x$-Achse liegen; ihre Verteilung längs dieses Streifens ist aber proportional dem $\int_a^b f_1(x)\, dx$, wenn $a$ bis $b$ ein beliebig variierendes Intervall ist. Denn jeder Punkt des Streifens entspricht auch einem Wert $x$, und deren Verteilung erhält sich unverändert, wenn der Verteilung der $y$ Beschränkungen auferlegt werden; dies ist eben die Forderung der Unabhängigkeit. Darum muß, wenn die Grenzen $c$ und $d$ festgehalten werden und $a$ und $b$ variieren,

This set-up cannot be derived from the existence of probability functions. It could be possible that every individual event follows exactly its probability function when it reoccurs, but that nevertheless the distribution of combinations is not given by the product. When throwing the two dice the same side could come up on both dice every time; this does not rule out that for each die every side comes up equally often but the frequency of every combination would not be given by the number $\frac{1}{36}$. The condition that there exists a probability function for every individual event is therefore not a sufficient justification for the law of composite probability.

However, the law can be justified if one assumes that there exists a probability function $f(x, y)$ for the combinations of $x$ and $y$, i. e., if one assumes that there exists a function that assigns to every combination of values $x$ from the interval $a$ to $b$ and values $y$ from the interval $c$ to $d$ its frequency given by the expression

$$\int_a^b \int_c^d f(x, y)\, dx\, dy$$

The existence of the integral implies that if the $x$-$y$-plane | is divided into 33
a large number of small rectangles, $\Delta x \times \Delta y$, and the corresponding columns reaching to the $f(x, y)$-surface representing the probabilities are divided into corresponding sets, then the sums within these sets are in the limit equally large. The proof is similar to the previous demonstration for functions of *one* variable. Each equally likely case corresponds to such a set, and the existence of a probability function suffices – as before – for these cases have equal frequency. Therefore the existence of such a function is a sufficient condition for the equal frequency of all combinations of throws of two dice.

The first condition requires this function to have the special feature that in any particular sequence of such combinations specified by this law, every set $x$ of values and every set $y$ of values are simultaneously governed by their particular probability function. Consider the experiment in which two different variables $x$ and $y$ are simultaneously physically realized in repeating trials; then a vast variety of different combinations of $x$ and $y$ will be observed. Let each observed combination $x, y$ be represented by a point in the $x$-$y$ plane. If the $y$ variable is now forced to remain within an interval $c$–$d$, then all these points will lie in a band parallel to the $x$-axis. The distribution along this band is proportional to $\int_a^b f_1(x)\, dx$ if the interval from $a$ to $b$ varies arbitrarily. Each point on the band corresponds to an $x$-value and their distribution remains unchanged when the distribution of $y$-values is restricted: this is the condition of independence. Hence, if the limits $c$ and $d$ are fixed while $a$ and $b$ vary arbitrarily, the following equality must hold:

$$\int_a^b \int_c^d f(x,y)\,dx\,dy = k' \int_a^b f_1(x)\,dx$$

sein. Das Gleiche gilt, wenn $a$ und $b$ festgehalten werden und $c$ und $d$ variieren; so daß

$$\int_a^b \int_c^d f(x,y)\,dx\,dy = k \int_a^b f_1(x)\,dx \int_c^d f_2(y)dy$$

34 werden muß. Diese spezielle Form für das Doppelintegral haben wir erhalten, indem wir die Verteilung auf einem bestimmten Wege vor sich gehen ließen. Nehmen wir nun an, daß die schließlich eintretende Verteilung unabhängig vom Wege ist - und dies haben wir eben dadurch ausgesprochen, daß wir sagten: es *existiert* eine die Verteilung der Kombinationen von $x$ und $y$ regelnde Funktion, die, wie auch der Weg des Versuchs sein mag, jeder Kombination von Werten $x$ aus dem Intervall $a$ bis $b$ und Werten $y$ aus dem Intervall $c$ bis $d$ durch das Doppelintegral ihre Häufigkeit zuschreibt – so gilt die abgeleitete Relation immer.

Da die Integrale durch Ausdehnung über den ganzen Bereich von $-\infty$ bis $+\infty$ gleich 1 werden sollen, folgt $k = 1$; und wegen der Unabhängigkeit der Funktionen $f_1(x)$ und $f_2(y)$ voneinander folgt

$$\int_a^b \int_c^d f(x,y)\,dx\,dy = k \int_a^b \int_c^d f_1(x) f_2(y)\,dx\,dy$$

und, da die Integranden positiv sind und die Genauigkeit für beliebig kleine Intervalle $b - a$, $d - c$ gilt:

$$f(x,y) = f_1(x) f_2(y)$$

Dieser Schluß gilt, wenn die Stetigkeit der Funktion nicht vorausgesetzt wird, nur bis auf eine Nullmenge von Punkten. Da diese zum Integral nichts beitragen, können die Funktionen hier verschieden sein. Gerade deshalb aber ist dies auch unwesentlich, denn es kommt für die Wahrscheinlichkeitsrechnung stets nur auf das Integral an.[7]

Die abgeleitete Relation läßt sich, wie man erkennt, ohne weiteres auf mehrere Werte $x, y, z \ldots$ übertragen, so daß

$$f(x,y,z\ldots) = f_1(x) f_2(y) f_3(z) \ldots$$

Demnach läßt sich das Theorem von der zusammengesetzten Wahrscheinlichkeit durch die beiden Voraussetzungen ersetzen, daß erstens eine Wahrscheinlichkeitsfunktion für die Verteilung der Kombinationen von $x, y, z \ldots$ existiert, und daß zweitens diese Verteilung gleichzeitig so erfolgt, daß sie jede einzelne
35 Wertreihe gemäß ihrer besonderen Wahrscheinlichkeitsfunktion regelt. | Die erste Voraussetzung kann nicht weggelassen werden, weil sonst die Unabhängigkeit der Verteilung vom Wege nicht gewährleistet ist.

7. Vgl. auch das auf Seite 42 über die Stetigkeit Gesagte.

$$\int_a^b \int_c^d f(x, y)\, dx\, dy = k' \int_a^b f_1(x)\, dx$$

The same is true when $a$ and $b$ are fixed while $c$ and $d$ vary; i. e.,

$$\int_a^b \int_c^d f(x, y)\, dx\, dy = k \int_a^b f_1(x)\, dx \int_c^d f_2(y) dy$$

We reached this particular form of the double integral by obtaining the distribution in a particular manner. If we now assume that the resulting distribution is independent of how it was obtained then the differentiated relation always holds. We explicitly stated the antecedent in claiming that there *exists* a function governing the distribution of combinations of $x$ and $y$, which, independent of the particular experiment, assigns the frequency given by the double integral to every combination of values $x$ in the interval $a$ to $b$ and values $y$ in the interval $c$ to $d$. 34

Since the integrals must be equal to 1 when expanded to the interval from $-\infty$ to $+\infty$, it follows that $k = 1$; and it follows due to the independence of the functions $f_1(x)$ and $f_2(y)$ that

$$\int_a^b \int_c^d f(x, y)\, dx\, dy = k \int_a^b \int_c^d f_1(x) f_2(y)\, dx\, dy$$

and since the integrands are positive and the equality holds for arbitrarily small intervals $b - a$, $d - c$, we have

$$f(x, y) = f_1(x) f_2(y)$$

If the continuity of the function cannot be assumed, then this result is true except for a null-set of points. Since these do not contribute anything to the integral, the functions may differ at these points. But this just shows that such points are irrelevant, since only the integral matters to probability calculus.[7]

It can be seen that the derived result can easily be extended to several variables $x, y, z \ldots$, giving

$$f(x, y, z \ldots) = f_1(x) f_2(y) f_3(z) \ldots$$

Consequently the theorem on composite probability can be replaced by the two conditions that first, a probability function for the distribution of combinations of $x, y, z \ldots$ exists, and second, that this distribution occurs simultaneously in such a way that each set of values for a particular variable occurs according to its particular probability function.[E6] | The first condition is required because 35

7. See also what is said about continuity on page 42 [99].

E6. Note that Reichenbach only considers joint distributions of independent events in this passage on composite probability.

Die Einführung der Wahrscheinlichkeitsfunktion mehrerer Veränderlicher ermöglicht nun eine genauere Fassung des Approximationsproblems, das bei der Definition der Wahrscheinlichkeitsfunktion auftritt. Jede Folge von Wiederholungen eines Vorgangs ist *endlich*, und es kann deshalb nie eine exakte Übereinstimmung mit der durch die Funktion definierten Verteilung erreicht werden, sondern nur eine Approximation. Auf diese Tatsache wurde bereits bei der Einführung der Wahrscheinlichkeitsfunktion hingewiesen. Die Art der Approximation soll nun zunächst so definiert werden: Wenn ich eine kleine Größe $\epsilon$ und ein $\Delta x$ vorgebe, so gibt es ein $N$ derart, daß in jedem Intervall die Abweichung[E2]

$$\left| \frac{h}{N} - \int_{x}^{x+\Delta x} \phi(x)\, dx \right| < \epsilon$$

wird. Hier ist $h$ die Anzahl der in das Intervall $x$ bis $x + \Delta x$ wirklich fallenden Werte.

Dies ist jedoch noch keine hinreichende Forderung. Denn eine Reihe von Wiederholungen, die diesem Gesetz genügt, könnte doch den Anforderungen, die man in der Wahrscheinlichkeitsrechnung in bezug auf die *Dispersion* an sie stellt, nicht genügen. Z.B. denke man sich eine Reihe von Würfen mit einem Würfel, in der die ersten beiden Würfe, dann der dritte und vierte Wurf, dann der fünfte und sechste Wurf usw. immer die gleiche Ziffer ergeben. Eine solche Reihe brauchte der eben geforderten Approximation nicht zu widersprechen, dagegen würde sie die von der Wahrscheinlichkeitsrechnung geforderte Dispersion nicht besitzen. Diese wird erst erreicht, wenn je zwei aufeinanderfolgende Würfe in ihrer Kombination auch noch durch Wahrscheinlichkeitsgesetze geregelt sind, d. h. wenn gleichzeitig eine Wahrscheinlichkeitsfunktion

$$\phi(x, x')$$

existiert, wo die Folge der Wiederholungen in Gruppen von je zwei aufeinan-
36 derfolgenden Elementen abgeteilt ist und $x$ und $x'$ | diese beiden Elemente bedeuten. Das läßt sich auf höhere Gruppen ausdehnen, und wir können nunmehr definieren:

Eine Wahrscheinlichkeitsverteilung ist dann vorhanden, wenn eine Folge von Funktionen

$$\begin{gathered} \phi(x) \\ \phi_1(x, x') \\ \cdots\cdots \\ \phi_r(x, x', x'', \ldots, x^{(r)}) \end{gathered}$$

existiert, die der Reihe der Wiederholungen in folgender Weise zugeordnet ist: wenn ich die beliebig große Zahl $r$, die beliebig kleine Größe $\epsilon$ und die Größen

E2. The original German had an error, the integral ranged from $-x$ to $x + \Delta x$.

otherwise one cannot ensure that the distribution is independent of how it was obtained.

The introduction of probability functions with several variables now allows for a more exact discussion of the problem of approximation that arises in the definition of a probability function. Every sequence of repetitions of a process is *finite* and therefore there can never be exact agreement, only an approximation to the distribution defined by the probability function. This fact was already mentioned when the probability function was introduced. Let us initially define the type of approximation in the following way: Given a small quantity $\epsilon$ and $\Delta x$, then there is an $N$ such that for every interval the difference

$$\left| \frac{h}{N} - \int_{x}^{x+\Delta x} \phi(x)\, dx \right| < \epsilon$$

where $h$ is the number of values actually falling within the interval $x$ to $x + \Delta x$.

However this is not yet a sufficient condition. A sequence of repetitions that satisfies this law may not satisfy the requirements that probability calculus sets with respect to *dispersion*. For example, consider a sequence of throws with a die in which the first two throws, then the third and fourth, and the fifth and sixth throw etc. always give the same value. Such a sequence need not contradict the approximation specified above, but it would not have the dispersion required by the probability calculus. This is only achieved if in addition the combination of any two consecutive throws is governed by the laws of probability, i. e., if at the same time a probability function

$$\phi(x, x')$$

exists, where the sequence of repetitions is divided into groups of two consecut-
ive elements, $x$ and $x'$. | This can be extended to higher groups and we can now 36
define:

A probability distribution is given when the sequence of functions

$$\begin{gathered} \phi(x) \\ \phi_1(x, x') \\ \ldots\ldots \\ \phi_r(x, x', x'', \ldots, x^{(r)}) \end{gathered}$$

exists and is assigned to the sequence of repetitions in the following manner: given the arbitrarily large number $r$, the arbitrarily small positive number $\epsilon$ and

$\Delta x, \ldots, \Delta x^{(r)}$ vorgebe, so gibt es ein $N$ derart, daß die Abweichung

$$\left| \frac{h^{(i)}}{N} - \int_x^{x+\Delta x} \cdots \int_{x^{(i)}}^{x^{(i)}+\Delta x^{(i)}} \phi_i(x, \ldots, x^{(i)})\, dx \ldots dx^{(i)} \right| < \epsilon$$

wird, und dies gleichzeitig durch dasselbe $N$ für alle Funktionen $\phi$ bis $\phi_r$. $h^{(i)}$ bedeutet wieder die Anzahl der wirklich in das Intervall fallenden Wiederholungen.

Dies ist die exakte Definition der Approximation. Die Wahrscheinlichkeitsrechnung fordert deshalb nicht nur die Existenz einer einzigen, sondern einer Folge von Wahrscheinlichkeitsfunktionen.

## 2.4 Die Fehlertheorie

Die Fehlertheorie geht von der Grundtatsache aus, daß es exakte Messungen nicht gibt; jede Beobachtung, die wir ausführen, wird dadurch in ihrer Genauigkeit gestört, daß unbekannte Einflüsse unser Meßinstrument verändern. Nun ist es zwar richtig, daß die bei jeder Wahrnehmung auftretende Empfindungsschwelle es uns schon aus rein psychologischen Gründen unmöglich macht, vollkommen exakt zu messen. Die Vergleichung zweier physischer Größen durch die Sinneswahrnehmung, auf die letzten Endes jede Messung hinausläuft, kann eben zwischen Größen keinen Unterschied mehr konstatieren, wenn diese nur
37 innerhalb der Wahr|nehmungsschwelle voneinander abweichen. Um den dadurch entstehenden Fehler möglichst klein zu machen, ist es das allgemeine Bestreben der Experimentatoren, Meßinstrumente herzustellen, die schon bei sehr kleinen Veränderungen des zu messenden Gegenstandes eine große Veränderung zeigen; man nennt solche Instrumente stark empfindlich. Aber es ist doch falsch, auf diese psychologischen Ursachen allein die prinzipielle Unexaktheit unserer Messungen zurückführen zu wollen. Denn noch eine andere Ursache macht uns die absolute Genauigkeit unmöglich; es hängt dies damit zusammen, daß es in der Natur keine „abgeschlossenen" Vorgänge gibt, wie bereits früher ausgeführt wurde. Jeder Naturvorgang unterliegt prinzipiell unendlich vielen äußeren Einflüssen; wir können immer nur einzelne herausgreifen und betrachten und müssen dann die anderen als verschwindend klein beiseite lassen. Der gemessene Wert ist stets die algebraische Summe unendlich vieler beeinflussender Größen. Wir rechnen aber so, als ob er durch endlich viele Faktoren gebildet ist, und berechnen dann daraus rückwarts die gewünschte Größe. Niemals messen wir direkt die Größe, auf die es ankommt, sondern stets eine derartige Funktion von ihr. Denken wir uns etwa eine einfache Länge durch mehrfaches Anlegen eines Meterstabes gemessen. Dann ist das zahlenmäßige Resultat nicht etwa ein unmittelbarer Ausdruck für die Länge. Es hängt vielmehr auch von den Winkeln ab, die die Stäbe beim Aneinanderlegen gebildet haben; von der Schwankung der Temperatur während der Messungszeit, von dem Einfluß der Strahlung, elektrischer Vorgänge usw. Die gefundene Zahlengröße setzt sich also in komplizierter Weise aus den Einzelwerten dieser unendlich vielen Größen zusammen. Aufgabe des Beobachters ist es nun, das Verhältnis, in

the quantities $\Delta x \ldots \Delta x^{(r)}$ then there is an $N$ such that the difference

$$\left| \frac{h^{(i)}}{N} - \int_{x}^{x+\Delta x} \cdots \int_{x^{(i)}}^{x^{(i)}+\Delta x^{(i)}} \phi_i(x \ldots x^{(i)})\, dx \ldots dx^{(i)} \right| < \epsilon$$

and this holds simultaneously for all functions $\phi$ to $\phi_r$. $h^{(i)}$ is again the number of instances that actually fall within the interval.

This is the exact definition of the approximation. Probability theory therefore requires the existence of a sequence of probability functions rather than a single such function.

## 2.4 Theory of Error

The theory of error starts from the basic fact that there is no such thing as exact measurement; every observation that we make will be affected in its precision by unknown effects interfering with our measuring instruments. Of course, for purely psychological reasons exact measurements are impossible because of sensory thresholds in perception. Every measurement amounts to the comparison of two physical quantities by the senses. It cannot be recognized if their
difference is less than the sensory threshold. | In order to minimize such er- 37
rors, the general aim of experimenters is to produce measuring instruments that show a large change even for very small changes in the object to be measured; such instruments are called highly sensitive. But it is incorrect to assign the inexactness of our measurement in principle to these psychological causes alone, since there is a further reason that makes absolute precision impossible. This has to do with the fact, as explained earlier, that there are no "closed" processes in nature. Every process in nature is in principle subject to an infinity of external influences; we can only ever pick out a few and consider them and leave the others aside as negligibly small. The measured value is always the algebraic sum of an infinity of influential quantities. However, we operate as if the measured value is formed by a sum of finite factors, and then calculate from it backwards to the desired quantity. We never directly measure the quantity that matters, but only such a function of it. Consider a simple length measured by the repeated application of a ruler. Then the numerical result is not a direct expression of the length. It also depends on the angles between the rod and the ruler, the changes in temperature during the time of measurement, the effect of radiation, electrical processes etc. The resulting numerical value is therefore composed in a complicated way of the individual values of these infinitely many quantities. The task of the observer is now to determine the ratio in which these quantities

dem diese Größen zusammenwirken, zu berechnen und von da aus auf die eine gewünschte Größe rückzuschließen. Bei flüchtigen Messungen wird man in unserem Beispiel sämtliche anderen Einflüsse vernachlässigen und den gefundenen Zahlenwert direkt gleich der Länge setzen; der vorsichtigere Beobachter wird dagegen bereits den Einfluß der Neigungswinkel der Stäbe berücksichtigen und den Wert der Länge erst durch Rechnung ermitteln. Da wir aber nicht in der Lage sind, sämtliche unendlichvielen Einflüsse in unserer Rechnung zum Ausdruck zu bringen, so ist uns deshalb eine exakte Messung prinzipiell versagt.
38 Selbst wenn unsere Wahrnehmung nicht an Emp|findungsschwellen gebunden wäre, würden wir also niemals durch Messung eine Größe exakt angeben können.[8]

Die Fehlertheorie ist also vor das Problem gestellt, unter einer Anzahl sich widersprechender Werte für ein und dieselbe Größe einen Wert besonders auszuzeichnen, der als geltender Wert angenommen werden soll. Nun ist es uns prinzipiell unmöglich, den wahren Wert der Größe dafür zu wählen, weil wir ihn nicht bestimmen können. Wir müssen deshalb einen anderen Wert wählen, der durch seine Stellung unter den übrigen Werten irgendwie bevorzugt erscheint. Die verschiedensten Werte sind dafür geeignet; es liegt besonders nahe, den häufigsten Wert auszuwählen. Dieser ist vor allen anderen dadurch ausgezeichnet, daß er in einer Reihe von Versuchen am häufigsten sich wiederholt; ich kann also, wenn ich ihn nur bestimmen kann, mit Gewißheit die Aussage machen - zwar nicht, daß jede einzelne Messung zu ihm führen wird, aber daß doch die größte Zahl der Messungen ihn ergibt. Die Bedeutung eines solchen Wertes besteht also darin, daß ich über sein Vorkommen in der Natur mit Gewißheit eine Aussage machen kann. Der Verzicht auf einen wahren Wert – und dieser würde allerdings mit Gewißheit gelten – ist also gar nicht so groß; ich kann diese Gewißheit ersetzen durch eine andere, die von einer Häufigkeit gilt. Natürlich muß dazu erst noch die Existenz eines solchen häufigsten Wertes bewiesen werden. Ehe wir jedoch auf diese Frage eingehen, soll das Verfahren, wie es tatsächlich in der Fehlertheorie angewandt wird, analysiert werden.

Wir wählen die Ableitung des Gaußschen Fehlergesetzes, wie sie durch die Hypothese der Elementarfehler gegeben wird. Dort wird der tatsächliche Messungsfehler entstanden gedacht aus einer Reihe einzelner Fehlerursachen, deren jede eine zahlenmäßige Abweichung von dem wirklichen Wert bedingt, so daß der Gesamtfehler durch Übereinanderlagerung aller dieser Elementarfehler entsteht. Nun wird auch die einzelne Fehlerursache nicht bei jedem Versuch denselben Betrag des Elementarfehlers ergeben, sondern auch dieser wird in sei-
39 ner Größe veränderlich sein. Denke ich | mir nun einen solchen Fehler innerhalb enger Grenzen variierend, so wird der aus ihnen allen zusammengesetzte Gesamtfehler bedeutend größere Schwankungen ausführen. Seine untere Gren-

8. Es muß das besonders betont werden, da z. B. A. Elsas, S. 588, es so hinstellt, als ob allein die Empfindungsschwelle der Grund für die Unexaktheit unserer Messungen ist. Er bestreitet übrigens an der gleichen Stelle den Zusammenhang zwischen Fehlertheorie und Wahrscheinlichkeitsrechnung; eine Behauptung, die durch das hier Folgende widerlegt wird.

interact and, from that, to infer the desired quantity. In rough measurements, one will neglect all other influences and simply set the numerical value found in measurement equal to the length; but the more careful observer will take the influence of the inclination of the rods into account and only establish the value of the length after some calculation. Since we are unable to express all infinitely many influences in our calculation we are denied exact measurement in principle. Even if our perception | were not bounded by the limits of our senses, we 38
would never be able to determine a quantity exactly by measurement.[8]

Hence the theory of error is confronted with the problem of distinguishing one particular value – taken to be correct – from among a number of contradictory values for one and the same quantity. In principle, it is impossible to select the true value for the quantity since we are unable to determine it. Hence, we have to choose a different value that is privileged in relation to alternative values. A variety of values would qualify; the choice of the most common value seems obvious. In comparison to all others, it is privileged since in a sequence of experiments it is repeated most often. Hence, if I can determine it, I can claim with certainty, not that every measurement will result in this value, but that the largest number of measurements will result in it. The meaning of such a value then is that I can make a claim with certainty about its occurrence in nature. Foregoing a true value – which would hold with certainty – is therefore no huge loss; I can replace this certainty with another about the frequency. Of course, the existence of such a most common value has to first be proven. But before we address this question, let us analyse how the process is actually carried out in the theory of error.

We consider the derivation of the Gaussian Error Law, as it is given by the hypothesis of elementary errors. Here the measurement error is thought to arise from a variety of different error sources, each of which results in a numerical divergence from the true value, so that the total error results from the combination of these elementary errors. It is not the case that each individual source of error contributes the same divergence in each trial. This quantity, too, will vary. If I | 39
think of this error as varying within certain bounds, then the resulting composite total error will have significantly higher variation. Its lower bound will be the

8. This has to be particularly emphasized, since for example A. Elsas, p. 588, presents the issue as if solely the limits of our senses are the reason for the imprecision of our senses. He also disputes in the same place the connection between the theory of error and probability theory; a claim that will be refuted by the following.

ze wird durch die Summe aller unteren Grenzen der Elementarfehler gebildet, seine obere Grenze durch die Summe der oberen. Diese extremen Werte sind also nur auf eine Weise zu erreichen; sie kommen nicht häufiger vor als die am seltensten vorkommende untere oder obere Grenze eines der Elementarfehler. Überhaupt sind die unteren Werte des Gesamtfehlergebietes nur durch die äußeren Werte der Elementarfehler darstellbar, während die inneren Werte viel mannigfacher entstehen können; wenn ein Elementarfehler extreme Werte annimmt, so kann ein anderer doch kleine Werte annehmen, so daß ihre Summe im Mittelgebiet des Gesamtfehlers liegt. So kommt es, daß selbst bei gleichmäßiger Verteilung der Häufigkeit in den Elementarfehlergebieten die Häufigkeit in dem Spielraum des Gesamtfehlers nicht überall gleich ist; sie wird vielmehr in den mittleren Gebieten ein Maximum haben und nach beiden Rändern hin abnehmen.

Um nun einen mathematischen Ausdruck für diese Verteilung zu finden, müssen verschiedene Voraussetzungen gemacht werden. Man erhält dann ein mathematisches Gesetz, das jedem Fehler seine Häufigkeit zuordnet. Nun sind je nach den Voraussetzungen verschiedene derartige Fehlergesetze denkbar; unter allen aber ist das Gaußsche durch die Einfachheit und Einsichtigkeit seiner Voraussetzungen ausgezeichnet. Eine Reihe von Untersuchungen sind über dies Problem angestellt worden; man kann als ihr Resultat bezeichnen, daß die Gaußsche Fehlerkurve auf folgenden Voraussetzungen beruht:[9]

1. Die Häufigkeit jedes Elementarfehlers ist durch eine Wahrscheinlichkeitsfunktion bestimmt.

2. Diese setzen sich nach dem Theorem der zusammengesetzten Wahrscheinlichkeit zusammen.

40 3. Es müssen sehr viele, voneinander unabhängige Elementarfehler gleicher Größenordnung zusammenwirken.

Die erste Prämisse verlangt, daß für jeden Elementarfehler eine Wahrscheinlichkeitsfunktion existiert; was darunter zu verstehen ist, ist in früherem ausgeführt. Doch werden über die Gestalt der Funktion keine Annahmen gemacht, jeder Einzelfehler darf eine ganz beliebige, seine Häufigkeit bestimmende Funktion besitzen. Nach dem durch die zweite Prämisse gegebenen Gesetz fügen sich diese Funktionen zusammen. Damit ist vorausgesetzt, daß die einzelnen Fehler unabhängig voneinander sind; ob dies jeweils erfüllt ist, muß die Erfahrung lehren. Die Forderung der Unabhängigkeit ist in Prämisse 3 nochmals ausgeprochen worden, damit in dieser alle Forderungen, deren Erfüllung empirisch konstatiert werden kann, zusammengefaßt sind. Aus dem Beweis, der früher für das Theorem der zusammengesetzten Wahrscheinlichkeit

9. Ausführliche Behandlungen des Problems geben Czuber, *Theorie der Beobachtungsfehler*, und Poincaré, *Probabilité*. Am deutlichsten treten die hier genannten drei Voraussetzungen hervor bei Hausdorff, der aus ihnen in kurzer und durchsichtiger Weise das Gaußsche Gesetz ableitet, so daß hier für die mathematische Seite des Problems direkt auf diese Abhandlung verwiesen werden kann.

sum of all the lower bounds of the elementary errors, its upper bound the sum of all the elementary upper bounds. Consequently there is only one way to attain these extreme bounds; they do not occur more often than the least frequent upper or lower bound of an elementary error. In general, the extreme values of the total error are only representable by the extreme values of the elementary errors, while intermediary values can arise in many diverse ways; if one elementary error assumes an extreme value, then another may assume small values, so that the sum will be in the central area of the total error. Hence it follows that even when the distribution of frequencies of values is uniform for the elementary errors, the frequency is not the same everywhere for the total error; rather it will have a maximum in the central areas and will decline towards both boundaries.

In order to obtain a mathematical expression for this distribution, a few assumptions have to be made. We will then have a mathematical law that will assign a frequency to every error. Depending on the assumptions, a variety of error laws are possible; but among all of them, the Gaussian error law is distinguished for its simplicity and the plausibility of its assumptions. Several analyses have been done on this problem; their conclusion is that the Gaussian error curve is based on the following assumptions:[9]

1. The frequency of the elementary error is determined by a probability function.

2. These are combined according to the theory of composite probability.

3. Many mutually independent elementary errors of the same order of mag- 40
nitude must work together [to produce the total error].

The first premise requires that there exists a probability function for every elementary error; what this means has been explained above. However, no assumptions are made about the form of the function: each elementary error may have an arbitrary function determining its frequency. The law given by the second premise determines how these functions are combined. This ensures that the individual errors are independent of each other; whether this is the case, experience must decide. The independence requirement is reiterated in the third premise in order to include together here all requirements that can be established empirically. It follows from the proof that was previously given for the theorem on composite probabilities that this includes the condition that there

9. A detailed treatment of this problem is given by Czuber, *Theorie der Beobachtungsfehler*, and Poincaré, *Probabilité*. Hausdorff makes the three assumptions mentioned here most explicit. He derives the Gaussian laws from them in a concise and clear manner, so that we can refer directly to his work for the mathematical aspect of this problem.

gegeben wurde, folgt, daß dies die Voraussetzung enthält, daß eine Funktion $f(x, y, \ldots)$ für die Verteilung der Fehlerkombinationen existiert; wir dürfen also die Prämisse 2 durch diese Voraussetzung ersetzen. Die Funktion nimmt den Wert $f_1(x)f_2(y)\ldots$, wenn $f_1(x)$,$f_2(y), \ldots$ die Wahrscheinlichkeitsfunktionen der Elementarfehler sind. Dies ist jedoch nicht die gesuchte Fehlerfunktion, sondern diese soll eine Häufigkeitsverteilung für die Summe aller Werte $x, y, \ldots$ darstellen. Das Problem, aus dem Produkt $f_1(x)f_2(y)\ldots$ die Funktion $\phi(s)$ abzuleiten, wo $s = x + y + \ldots$, ist das mathematische Problem des Fehlergesetzes. In der Abhandlung von Hausdorff ist dieses allgemein behandelt worden, es ist ein allgemeines Verfahren zur Gewinnung des Gesetzes des Gesamtfehlers angegeben worden, in dem das Gaußsche Gesetz als Näherung für einen besonderen Fall erscheint. Dieser besondere Fall ist dann verwirklicht, wenn sehr viele, voneinander unabhängige Elementarfehler gleicher Größenordnung zusammenwirken. Das ist jedoch durchaus nicht immer erfüllt. Hausdorff gibt Beispiele für andere Fälle an. Das Gaußsche Gesetz ist deshalb keinesfalls als Prinzip der Wahrscheinlichkeitsrechnung anzusprechen; sondern es bedeutet eine ganz spezielle Form, die für den Fall der Prämisse 3 angenommen wird.

Das Gaußsche Gesetz ist dadurch ausgezeichnet, daß das arithmethische
41 Mittel der wahrscheinlichste Wert der Messungen wird, | d. h. daß es dem Maximalwert der Kurve $\phi(s)$ entspricht.[10] Es wird damit das arithmetische Mittel als derjenige Wert ausgezeichnet, der (genauer: dessen Nähe) von allen Messungen am häufigsten getroffen wird. Damit erscheint das vielbenutzte Verfahren, unter allen Messungen das arithmetische Mittel als wahrscheinlichsten Wert auszusuchen, gerechtfertigt. Wir erkennen jedoch, daß es keineswegs überall zulässig ist, dem arithmetischen Mittel diese Sonderstellung zu geben. Die erste und zweite Voraussetzung allerdings müssen wir allgemein für jede Fehlerverteilung machen; die dritte jedoch nur in besonderen Fällen. Zwar sind unsere modernen, komplizierten Meßinstrumente meist so gebaut, daß in ihnen viele Fehler gleicher Größenordnung zusammenwirken; aber es wird stets Fälle anderer Natur geben, in denen man also nicht das arithmetische Mittel auswählen darf. Das erscheint befremdend, weil man geneigt ist, in der Bevorzugung des arithmeti-

10. Die Gaußsche Funktion

$$\frac{h}{\sqrt{\pi}} e^{-h^2(s-a)^2}$$

erreicht ihr Maximum für $s = a$. Das Mittel der Werte $s$ ist gegeben durch

$$\int s\phi(s)ds$$

in der Tat ergibt sich

$$\frac{h}{\sqrt{\pi}} \int_{-\infty}^{+\infty} se^{-h^2(s-a)^2} ds = a$$

Da die drei Prämissen die Gaußsche Funktion eindeutig bestimmen, so folgt auch notwendig diese Konsequenz.

exists a function $f(x, y \ldots)$ for the distribution of combinations of errors; hence we may replace premise 2 by this condition. The function will assume the value $f_1(x)f_2(y)\ldots$, if $f_1(x), f_2(y)\ldots$ are the probability functions of the elementary errors. However, this is not the desired error function, instead it describes a distribution of frequencies for the sum of all values $x, y \ldots$. The mathematical problem of the error law is to derive the function $\phi(s)$ from the product $f_1(x)f_2(y)\ldots$, where $s = x+y+\ldots$. In the work by Hausdorff this has been analyzed generally: he presents a general method for obtaining the law of total error and the Gaussian law appears as an approximation for a special case. This special case occurs if many mutually independent elementary errors of the same order of magnitude operate. Obviously this is not always the case and Hausdorff gives other examples. Hence, one cannot speak of the Gaussian law as the principle of the probability calculus; rather it is a special case when premise 3 is assumed.

The Gaussian law is distinguished in the sense that the arithmetic mean becomes the most probable value of the measurements, | i. e., it coincides with the 41
maximum of the curve $\phi(s)$.[10] The arithmetic mean is therefore determined as the value that (to be exact: whose proximity) is hit the most often by all measurements. The widely used method of choosing the most likely value as the arithmetic mean would thereby seem to be justified. However, we recognize that it is by no means acceptable to assign this special status to the arithmetic mean in all cases. We have to require the first and second condition generally for all error distributions; the third, however, only in special circumstances. Our modern, complicated measuring instruments are generally constructed so that many errors of the same magnitude operate, but there will always be cases of a different nature in which one must not select the arithmetical mean. This seems unintuitive, since one tends to see the preference for the arithmetic mean almost as

10. The Gaussian function

$$\frac{h}{\sqrt{\pi}} e^{-h^2(s-a)^2}$$

attains its maximum for $s = a$. The mean of the values $s$ is given by

$$\int s\phi(s)ds$$

and does in fact result from

$$\frac{h}{\sqrt{\pi}} \int_{-\infty}^{+\infty} s e^{-h^2(s-a)^2} ds = a$$

Since the three premisses determine the Gaussian function uniquely, this consequence also necessarily follows.

schen Mittels geradezu das apriorische Prinzip der Fehlertheorie zu sehen. Es leuchtet ein, anzunehmen, daß die Gesamtabweichung vom wahren Messungswert nach der einen Seite ebenso groß ist wie nach der anderen. Man ist auch geradezu von dieser Annahme ausgegangen und hat dann unter Zuhilfenahme anderer Prämissen das Gaußsche Gesetz abgeleitet. Dies ist der ursprüngliche Weg von Gauss gewesen. Während in all diesen Ableitungen gewöhnlich die Stetigkeit und Differenzierbarkeit der Wahrscheinlichkeitsfunktion vorausgesetzt wird, hat F. Bernstein eine Ableitung gegeben, die sich mit der Integrierbarkeit begnügt.[11] Da diese Eigenschaft weniger von einer Funktion fordert als die der
42 | Stetigkeit und Differenzierbarkeit, ist die Ableitung mathematisch von Bedeutung; für das philosophische Problem hat es jedoch geringes Interesse, ob man der Wahrscheinlichkeitsfunktion außer der Integrierbarkeit auch noch Stetigkeit und Differenzierbarkeit zuerkennt oder nicht. Denn wenn man, wie wir es getan haben, nur die Integrierbarkeit der Funktion fordert, so läßt sich jederzeit eine wenigstens abteilungsweise stetige Funktion angeben, die mit beliebiger Näherung sich an die integrierbare Funktion anschließt. So stellt z. B. der von uns bei der Definition der Wahrscheinlichkeitsfunktion erwähnte Polygonzug eine solche abteilungsweise stetige und differenzierbare Kurve dar, die mit beliebiger Näherung an Stelle der anderen gesetzt werden kann. Man kann deshalb stets in jedem speziellen Fall anstatt der integrierbaren Funktion eine von diesen Näherungsfunktionen nehmen und aus ihr, unter Voraussetzung der Stetigkeit und Differenzierbarkeit die gewünschten Resultate ableiten; die zahlenmäßige Auswertung der Resultate wird dann mit einer entsprechenden Näherung von dem wirklich gemessenen Verhalten gelten. Die Annäherung in den Resultaten wird sich also ebenfalls zahlenmäßig beliebig klein machen lassen. Poincaré hat ferner darauf hingewiesen, daß die von Gauss vorausgesetzte Wahrscheinlichkeitsfunktion die Annahme einschließt, daß die Wahrscheinlichkeit nur von der Größe des Fehlers, nicht auch von der Größe des Messungswertes abhängt.[12] Er hat für Fälle, in denen dies nicht erfüllt ist, eine besondere Wahrscheinlichkeitsfunktion abgeleitet.[13] Gerade die Poincaréschen Untersuchungen zeigen, daß dem arithmetischen Mittel durchaus nicht a priori die Stellung des wahrscheinlichsten Wertes zukommt, sondern daß man ihm diese Bedeutung nur in

11. Seine Prämissen sind: a) Das arithmetische Mittel ist der wahrscheinlichste Wert der Beobachtungen, b) Es existiert eine Fehlerfunktion, die die Eigenschaft der Integrierbarkeit besitzt.

12. Poincaré, *Probabilité*, p. 173.

13. Es ist immer dann erfüllt, wenn die Schwankungen des Messungswertes klein sind gegenüber den Instrumentalfehlern, so daß der Messungswert als konstant betrachtet werden kann. In diesem Falle kann man auch von der Bestimmung des „wahren" Wertes reden, wenn man darauf ausgeht denjenigen Störungseinfluß, der lediglich von den Instrumentalfehlern herrührt, zu eliminieren. Dazu muß man den „systematischen Fehlern" (die Konstante $a$ der Funktion) bestimmen; dies ist mit Hilfe eines sehr viel feineren Meßinstruments möglich. Solche Bestimmung der „Aparatkorrektion" geschieht häufig bei Sextanten, Thermometern, Universalen usw. In besonderen Fällen darf man annehmen, daß $a = 0$ ist; dies läßt sich gelegentlich aus der Symmetrie der Streuungsbedingungen schließen, gelegentlich auch aus anderen Nebenumständen. Apriori aber läßt sich über $a$ nichts sagen.

an a priori principle of the theory of error. It seems plausible to assume that the total deviation from the true measurement value is just as large in one direction as in the other. The assumption was often made and the Gaussian law then derived with the help of other premisses. This was Gauss's original approach. While all these derivations generally assume the continuity and differentiability of the probability function, F. Bernstein provided a derivation that only required its integrability.[11] Since this property is a weaker requirement of the function
than | continuity and differentiability, this derivation is of mathematical import- 42
ance; however, for the philosophical problem it is of less interest whether one requires continuity and differentiability in addition to integrability of the probability function. Since if, as we have done, one only requires integrability, then one can always specify at least a partially continuous function that approximates with arbitrary precision an integrable function. Hence, for example, the polygon sequence we presented in the context of the definition of the probability function is such a partially continuous and differentiable curve that can replace the true function to arbitrary approximation. Consequently, in any particular case one can use one of these approximate functions instead of the integrable function and derive the desired results from it under the assumption of continuity and differentiability. The numerical conclusions of the results will hold to within the corresponding approximation of the true measured behavior. So the error of the approximation can also be made arbitrarily small. Poincaré further pointed out that the Gaussian probability function requires the assumption that the probability only depends on the size of the error but not on the size of the value to be measured.[12] For cases in which this does not hold he derived a special probability function.[13] Poincaré's analyses in particular show that the arithmetical mean certainly does not have a priori status as the most probable value, but

11. His premises are: a) The arithmetic mean is the most probable value in the observations. b) There exists an error function that is integrable.

12. Poincaré, *Probabilité*, p. 173.

13. [That the probability only depends on the size of the error but not on the size of the value to be measured] always holds if the variance of the measured value is small compared to instrument errors, so that the measured value can be considered constant. In this case one can also speak of determining the "true" value if one assumes that the influence that derives from instrument errors alone is eliminated. To do so, one has to determine the "systematic error" (the constant $a$ in the function); this can be done with a much more accurate measuring instrument. Such analyses of the "instrument correction" are often done for sextants, thermometers, universals etc. In special cases one can assume that $a = 0$; this can sometimes be concluded from the symmetry of the scattering conditions, sometimes also from other side conditions. But nothing can be said about $a$ a priori.

43 beson|deren Fällen einräumen darf, in denen besondere empirische Gründe dafür sprechen.

Die Prämisse 3 charakterisiert derartige Fälle; ob sie jeweils erfüllt ist, kann nur die Erfahrung lehren. Weiß man nichts darüber, so erscheint die Anwendung des Gaußschen Gesetzes gezwungen; richtiger wäre dann eine Kurve, die sich den Beobachtungen besser anschmiegt. Man könnte dann in der Weise verfahren, daß man den Spielraum des Fehlers in kleine gleiche Intervalle teilt, zählt, wieviel Werte in jedes Intervall gefallen sind und diese Häufigkeiten als Ordinaten über den entsprechenden Fehlerintervallen abträgt. Zieht man dann durch die Endpunkte eine Kurve, so bedeutet diese eine erste Näherung an das Fehlergesetz. Ein mathematisches Verfahren, diese Kurve zu bestimmen, hat Bertrand angegeben.[14] In der Figur ist diese Kurve durch die ausgezogene Linie

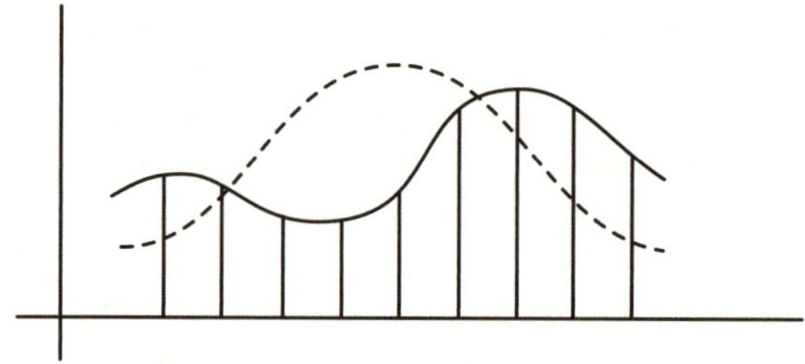

Abbildung 3.

dargestellt; würde man auf diesen Fall das Gaußsche Gesetz anwenden, so erhielte man die punktierte Linie, die offensichtlich als eine gewisse Vergewaltigung der Messungsresultate erscheint. Man würde mit der Wahl dieser Kurve behaupten, daß die Streuungsbedingungen des Messungsverfahrens so groß sind, daß die Zahl der Messungen für eine gute Annäherung noch zu klein ist; erst bei wei-
44 terer Wiederholung der Messung, | so müßte man sich rechtfertigen, wird eine Annäherung der wirklichen Verteilung an die Gaußsche Kurve erreicht. Wenn man gute Gründe für die Annahme hat, daß Prämisse 3 erfüllt ist, erscheint dies auch gerechtfertigt.

Es ergibt sich aus diesen Darlegungen, daß das Gaußsche Fehlergesetz eine ganz spezielle Form ist, die durchaus nicht den allgemeinen Grundsatz der Feh-

14. Er geht dazu von dem Grundsatz aus, daß das arithmetische Mittel nicht der wahrschein*lichste*, sondern der wahrschein*liche* Wert werden soll; d. h. daß

$$\frac{\sum x}{n} = \int_{x_{min}}^{x_{max}} x\phi(x)\,dx$$

werden soll. ($x$ Messungswert, $n$ deren Anzahl, $x_{min}$ und $x_{max}$ der kleinste resp. größte Messungswert.) Für eine Verteilung, die dem Gaußschen Gesetz schon folgt, ist dies natürlich auch erfüllt; wenn aber die Verteilung anders geartet ist, etwa wie in der Figur, so würde die Anwendung des Gaußschen Gesetzes nur eine sehr grobe Näherung ergeben. Vgl. Poincaré, Probabilité, p. 173.

rather that this meaning can only be assigned to it | under special circumstances 43
where there are empirical reasons to believe as much.

Premise 3 characterizes such cases; whether it obtains only experience can tell. If one has no prior knowledge, then one appears forced to use the Gaussian law; better would be a curve that follows the observations more closely. One could proceed by dividing the error space into small equal intervals, count how many values fell within each interval and plot these frequencies as ordinates above the corresponding error intervals. The curve defined by these points is a first approximation to the error law. Bertrand describes a mathematical procedure for determining this curve.[14] In the figure this curve is shown with a solid

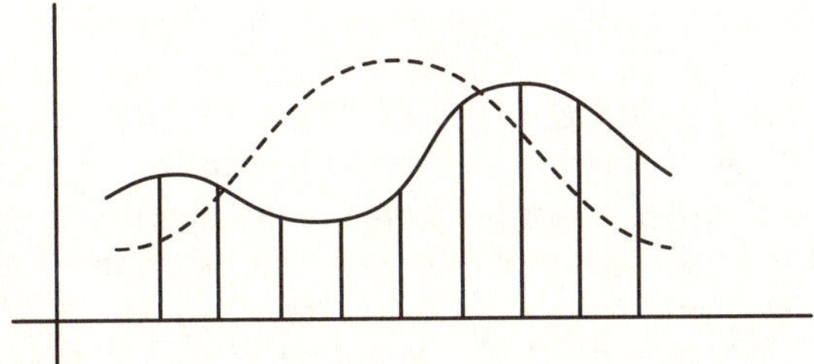

Figure 3.

line; if one were to use the Gaussian law in this case, then one would get the dotted line, which in this case obviously appears to be a brutal distortion of the measurement results. In selecting this curve one would be claiming that the scattering conditions of the measurement procedure are so large that the number of measurements is still too small for a good approximation; the justification would
have to be that only after further repetitions of the measurement | will we achieve 44
an approximation of the true distribution to the Gaussian curve. If one has good reasons for the assumption that premise 3 holds, then this seems justified.

It follows from this discussion that the Gaussian error law has a very special form that does not in general provide the fundamental principle of the theory of

14. He assumes the basic principle not that the arithmetic mean will be the *most* probable but rather the *expected* value; i. e., that

$$\frac{\sum x}{n} = \int_{x_{min}}^{x_{max}} x\phi(x)\, dx$$

($x$ is the measured value, $n$ the number of such values, $x_{min}$ and $x_{max}$ the smallest and largest measured value, respectively.) Of course, this also holds for the distribution that follows the Gaussian law; however, if the distribution is of a different sort, for example as in the figure, then the usage of the Gaussian Law would only provide a very rough approximation. See Poincaré, *Probabilité*, p. 173.

lertheorie darstellt. Ebenso ist die sogenannte Hypothese des arithmetischen Mittels, das ist die Annahme, daß das arithmetische Mittel der wahrscheinlichste Wert sei, nicht der allgemeine Grundsatz der Fehlerrechnung, sondern nur für besondere Fälle gültig. Allein die erste und zweite Prämisse enthalten Gedanken, die ganz allgemein für die Fehlertheorie gültig sind; kein Fehlergesetz kann ohne die Annahme auskommen, daß überhaupt eine Wahrscheinlichkeitsfunktion existiert, möge diese nun eine Funktion von einer oder mehreren Veränderlichen sein. Und man erkennt, daß dieser Gedanke kein anderer ist als der, der früher als implizite Voraussetzung der Theorie der Wahrscheinlichkeitsmaschine und der Glücksspiele aufgedeckt worden war.

Diese seltsame Übereinstimmung in zwei anscheinend so getrennten Gebieten wird dem nicht befremdend erscheinen, der die im früheren beschriebene und analysierte ideale Wahrscheinlichkeitsmaschine einmal von der anderen Seite betrachtet hat. Eigentlich war in der Wahl dieses Apparats zum Idealfall bereits die einheitliche Auffassung dieser Gebiete vollzogen. Denn jene Maschine unterscheidet sich nicht wesentlich von einem Apparat zur Messung der Fallzeiten des Geschosses. Ein solcher würde die Fallzeiten mit einer Uhr messen, d. h. mit den Umlaufszeiten eines Rades vergleichen. Hier ist an Stelle des Rades das rollende Band getreten. Die kleinen Differenzen in den Fallzeiten waren es, die die Anwendung der Wahrscheinlichkeitsbetrachtung überhaupt erst ermöglichten. Über ihre Verteilung in der Zeit mußte eine eigentümliche Annahme gemacht werden. Und dieselben kleinen Schwankungen der Fallzeit sind der Gegenstand der Fehlertheorie.[15] Denn zur exakten Messung der Fallzeit würde man die Messung häufig wiederholen und in der folgenden fehlertheoretischen
45 Behandlung | die Voraussetzung der Existenz einer Wahrscheinlichkeitsfunktion machen. Daß es beide Male die gleiche eigentümliche Struktur in der Verteilung der Fallzeitwerte ist, die wir voraussetzen, das ist ein wichtiges Resultat der bisher durchgeführten Untersuchungen. Wenn sich ein Widerspruch in den Voraussetzungen ergeben hätte, müßten wir eine der beiden Theorien als falsch verwerfen. Die Analyse hat jedoch die Identität gelehrt.

Jeder Fall, wo ein physikalischer Vorgang häufig wiederholt wird, der also in der Physik nach der Fehlertheorie beurteilt wird, stellt eine für Wahrscheinlichkeitsexperimente geeignete Anordnung dar. Nur unterliegt er nicht immer dem Gaußschen Gesetz. Beim Roulettespiel, beim Würfeln z. B. sind die Bedingungen des Gaußschen Gesetzes nicht erfüllt; dort überwiegt der eine Fehler, der in den Schwankungen der Stoßkraft des Armes, der Feder gegeben ist, so sehr alle anderen, daß Prämisse 3 nicht mehr anwendbar ist. Hier ist die Wahrscheinlichkeitsfunktion nicht eine eingipfelige Kurve, sondern vielfach gewunden; aber das eine bleibt doch erfüllt, daß überhaupt eine Wahrscheinlichkeitsfunktion *existiert.* Hier liegt das philosophische Problem. Woher weiß ich, daß es solch eine Funktion gibt?

15. Für den beschriebenen Apparat sollen die Schwankungen der Fallzeit groß sein gegenüber den übrigen Schwankungen des Apparates.

error. Similarly, the so-called hypothesis of the arithmetic mean, i. e., that the arithmetic mean is the most probable value, is not the general, basic principle of the theory of error, but only holds under special circumstances. Only the first and second premises contain ideas that hold generally in the theory of error; no law of error can be established without the assumption that there exists a probability function at all, no matter whether it is a function of one or several variables. And one recognizes that this thought is no different from the one that we found earlier to be an implicit condition of the theory of the probability machine and of games of chance.

This strange agreement in two such seemingly separate areas will not appear surprising to those who have considered the probability machine we previously described and analyzed it from a different perspective. The unified understanding of these areas had basically already been achieved in the choice of this apparatus as an ideal case, since this machine does not differ significantly from an apparatus to measure the drop times of the projectile. Such a machine would measure the drop times with a clock, i. e., by comparison with the revolution times of a wheel. In place of the wheel, we have the moving tape. The small differences in drop times were what allowed for the analysis in terms of probabilities in the first place. A peculiar assumption had to be made about their distribution in time. And the same small variations of the drop time are the subject of the theory of error.[15] In order to obtain an exact measure of the drop time one would repeat the measurement several times and make | the assumption of the 45
existence of a probability function in the subsequent error-theoretical analysis. The most important result of the analysis so far is that in both cases we assume the same peculiar structure in the distribution of the drop times. If we had found a contradiction in one of the conditions, then we would have to reject one of the two theories as false. However, the analysis showed their identity.

Every case where a physical process is repeated often, and is to be analysed in physics according to the theory of error, presents a set-up appropriate for probability experiments. However, the Gaussian law may not apply. In a game of roulette or when throwing dice, the conditions for the Gaussian law are not satisfied; in this case the error resulting from the variation in the thrust of the arm[E7] dominates all other errors so strongly that premise 3 is not applicable. Here the probability function is not a unimodal curve, but has several maxima; but it does hold that a probability function *exists*. Here we have the philosophical problem. How do I know that there is such a function?

15. In the case of the machine, the variations of the drop time are supposed to be large compared with any other variation of the machine.

E7. The German contains a reference to a spring here, but it is not clear what it is referring to – a spring in the arm or a spring in the probability machine?

Für alle anderen Bedingungen war gezeigt worden, daß sie aus der Erfahrung abgeleitete Sätze sind; von Fall zu Fall mußte und konnte entschieden werden, ob sie erfüllt sind. Es handelte sich in ihnen stets um Messung bzw. Abschätzung von Größen. Für diese eine Bedingung aber versagt die Erfahrung. Eine endliche Zahl von Vorgängen kann ich beobachten und über ihre Verteilung etwas aussagen. Dieser Grundsatz aber sagt gerade etwas über die Wiederholung von Vorgängen; er behauptet, daß, wie die Verteilung auch zu Anfang sei, in ihren Häufigkeitszahlen doch schließlich eine Annäherung stattfinden müsse an eine integrierbare Funktion, und das sowohl für die Wiederholung eines einzelnen Vorgangs, als auch für die Wiederholung von Kombinationen. Die spezielle Form dieser Funktion muß die Erfahrung bestimmen; aber daß sich überhaupt eine derartige Aussage machen läßt über *beliebig viele Fälle*, das kann keine Erfahrung lehren. Denn nur über eine endliche Zahl von Fällen kann die Beobachtung Aufschluß geben. Es ist hier wie beim Kausalgesetz; sein spezieller Inhalt wird jeweils empirisch gewonnen, aber daß das im Einzelfall Beobachtete nun
46 *allgemein* gilt, daß von nun ab | jeder *beliebige* Fall dieser Gesetzlichkeit unterliegen soll, läßt sich durch keinerlei Beobachtung feststellen. Dennoch ist die hier behauptete Gesetzlichkeit eine ganz andere als die kausale. Denn sie behauptet ja ein Gesetz gerade über Vorgänge, die *nicht* kausal zusammenhängen. Überall dort, wo wir mit Hilfe des Kausalgesetzes nichts mehr über das Naturgeschehen ausmachen können, greift das Prinzip der Wahrscheinlichkeitsfunktion ein; dieses ist jetzt als eigentlicher Inhalt des Wahrscheinlichkeitsgesetzes nachgewiesen.

Wenn dieses Gesetz aber nicht aus der Erfahrung stammt, wo liegt dann die Rechtfertigung zu seinem Gebrauch? Diese Frage soll in der nun folgenden Deduktion beantwortet werden.

## 47 3. Deduktion des Wahrscheinlichkeitsprinzips

Wenn es möglich war, zu zeigen, daß ein Satz die Grundvoraussetzung ausgedehnter Wissensgebiete bildet, so ist für den Beweis seiner Gültigkeit von der Erfahrung bereits viel geleistet. Denn er darf alsdann keinen geringeren Anspruch auf Geltung machen als eben jene Wissensgebiete selbst; und mit ihrer Anerkennung wäre seine Gültigkeit zugegeben. Deshalb durfte Kant den Satz, daß die Axiome der Geometrie synthetische Urteile apriori seien, mit der gleichen Gewißheit aussprechen, die für die Geltung der Geometrie bestand; und da jedes vernünftige Wesen schlechterdings deren apodiktische Gewißheit zugeben mußte und zugab, so war mit der Aufzeigung, daß dieser Satz die notwendige Voraussetzung der Geometrie war, bereits der Beweis für seine Geltung angetreten. Für den hier zu besprechenden Fall ist jedoch die Sachlage ein wenig verändert. Denn die Wissensgebiete, um die es sich hier handelt, die Fehlertheorie und die Theorie der Zufallsspiele, genießen nicht jene unbestrittene Anerkennung, deren sich die Geometrie erfreuen kann. Die einen sehen in ihnen eine rein mathematische Spielerei, die keinerlei Beziehung auf die Welt der Dinge

All other conditions were shown to be derived from experience; in each case one had to and could decide whether or not they were true. We were always dealing with the measurement or estimation of quantities. But for this one condition, experience fails us. I can observe a finite number of processes and can say something about their distribution. But this principle makes a claim about the repetition of processes; it claims that no matter what the distribution is initially, the frequencies eventually have to approach an integrable function, both for the repetition of an individual process and for the repetition of combinations. The particular form of this function has to be determined empirically; but, since observation can only be informative about a finite number of cases, there is no experience that shows it is possible to make such a claim about *arbitrarily many repetitions*. We have a situation similar to the law of causality; each time its particular content is determined empirically, but it cannot be determined by observation that what is observed in the individual instance holds *generally*, | that from 46
now on every *arbitrary* instance will follow this law. Nevertheless, the lawfulness postulated here is very different from the causal one. After all it describes a law about processes that are *not* causally related. Anywhere that we are unable to make any progress on events in nature with the help of the causal law, the principle of the probability function applies. This has now been shown to be the actual content of the law of probability.

But if this law is not derived from experience, what is the justification for its use? This question shall be answered in the following deduction.

## 3. Deduction of the Principle of Probability 47

If it is possible to show that a principle forms the basic assumption for wide areas of knowledge, then a great deal has already been done towards the empirical proof of its validity. Since then it may make no lesser claim to its validity than these areas of knowledge make for themselves and, in recognizing them, one would admit the validity of the principle. This is why Kant could make the claim that the axioms of geometry are synthetic a priori judgments with the same certainty that held for the validity of geometry; and since every rational being had to and did admit to its apodictical certainty, then showing that this statement formed the necessary condition for geometry was already part of the proof of its validity. In the case described here, the situation is a little different. The areas of knowledge that we are considering here, the theory of error and the theory of games of chance, do not enjoy such undisputed respect as does geometry. Some consider [these subjects] to be a purely mathematical game that bear no relation to the world of things; others want to see in them the last safe-haven which

hat; die anderen wollen in ihnen die letzte Zuflucht sehen, die sich die menschliche Unwissenheit gebaut hat, um doch wenigstens noch irgendetwas behaupten zu können. Zwar bin ich der Ansicht, daß jeder, der in diesen Gebieten arbeitet, sich auf ihre Geltung mit Bestimmtheit verläßt, und es ist ja auch ein psycho-
48 logisches Faktum, daß die Über|zeugung von der Gültigkeit einer Wissenschaft nur durch längere Vertrautheit mit ihr erworben wird – ohne daß durch diese Behauptung diese Überzeugung selbst zu einer psychologischen Erkenntnis gestempelt werden soll. Denn es handelt sich hier letzterdings um eine Einsicht, um eine unmittelbare Erkenntnis, daß ihren Sätzen Geltung zukommt, und gerade dieser spontane Akt ist letzten Endes nicht weiter analysierbar und auch nicht weiter begründbar; deshalb kann er ebensowohl am Anfang wie am Ende aller begrifflichen Deduktion stehen. Trotzdem kann auf eine Konstruktion des begrifflichen Zusammenhangs nicht verzichtet werden. Denn sie erst gibt dem vorher nur durch reine Einsicht Erfaßten seinen letzten, inneren Halt, indem sie es in notwendigen Zusammenhang bringt mit den übrigen Grundsätzen der Erkenntnis; sie lehrt erst die Einheit alles Erkennens und gibt umgekehrt dadurch den durch Einsicht gewonnenen Resultaten ihren letzten transzendentalen Beweis, indem sie sie als von der Einheit der Erkenntnis notwendig gefordert hinstellt. So soll auch hier der Zusammenhang des einen metaphysischen Prinzips mit den wesentlichen Grundsätzen der Erkenntnis aufgezeigt werden. Dadurch wird die Gewißheit des Prinzips nicht allein auf die des vielfach umstrittenen Wissensgebietes der Fehlertheorie und der Zufallsspiele gestützt, sondern auf die der ganzen Erkenntnis überhaupt, und gleichzeitig dadurch der Streit über die Gewißheit dieser Gebiete dahin entschieden, daß mit der Anerkennung einer Naturerkenntnis auch die Anerkennung der Wahrscheinlichkeitstheorie gegeben ist.

Die Urteile von zeit-räumlichen Gegenständen lassen sich in zwei Klassen einteilen. Einmal kann das Urteil durch reine Anschauung und Begriffe des reinen Denkens allein gegeben sein, dann ist es ein mathematisches Urteil. Oder sein Inhalt ist durch empirische Anschauung bestimmt, nur durch unmittelbare Wahrnehmung (oder durch die Vorstellung einer solchen) gegeben, dann ist es ein physikalisches Urteil.

Der anschauliche Charakter des physikalischen Urteils soll an einem Beispiel dargetan werden. Betrachten wir den Satz, daß ein von einem Turm herabfallender Stein nach einer durch Galileis Formel genau bestimmten Zeit den Bo-
49 den berührt. Was | mit dem Begriff Stein gemeint ist, ist letzten Endes nur durch empirische Anschauung zu erfassen; man muß schon hinsehen auf einen wirklichen Stein oder die auf solche Weise erzeugte Wahrnehmung im Gedächtnis reproduzieren, um zu wissen, was der Begriff Stein besagen soll. Ebenso ist es mit dem Vorgang des Fallens und mit den übrigen in dem Urteil vorkommenden Begriffen. Auch die Größe der Zeit, von der die Rede ist, kann nur durch empirische Anschauung gegeben werden. Zwar muß die Zeit als reine Anschauungsform von aller empirisch wahrgenommenen Zeitgröße unterschieden werden; aber hier wird eben eine ganz *bestimmte* Zeitgröße behauptet. Der Sinn der Aussage geht dahin, daß die Gleichheit der Fallzeit mit einer andern, ebenfalls nur

human ignorance has built itself in order to at least be able to claim something. I hold the view that anyone who works in this subject is certain of its validity. It is also a psychological fact that the belief | in the validity of a science is only 48
obtained after long familiarization – without wishing by this claim to make this conviction itself a psychological finding. In the end we are dealing with an insight, an unmediated recognition that these claims are valid, and this spontaneous act is ultimately not further analyzable and cannot be justified; hence it can be made just as well at the beginning as at the end of any conceptual deduction. Nevertheless we cannot forego a construction of the conceptual context, since, by placing the insight within the context of other basic principles of knowledge, the construction provides the principle with its final inner support. Only the construction teaches the unity of all understanding and reciprocally provides the transcendental proof of the results gained by insight, by presenting them as necessarily required for the unity of understanding. In the same way we intend to show here the connection between this single metaphysical principle and the essential basic principles of knowledge. Thereby, the certainty of the principle is not solely supported by the disputed areas of knowledge, the theory of error and games of chance, but also by the certainty of knowledge in general. At the same time, the dispute over the certainty of these domains will be decided in the sense that an acceptance of knowledge of nature includes an acceptance of probability theory.

Judgments about spatio-temporal objects can be divided into two classes. On the one hand, the judgment can be given by pure intuition and concepts of pure thought alone; then it is a mathematical judgment. Or, on the other hand, its content is determined by empirical considerations, by direct observation (or by the imagination of such); then it is a physical judgment.

Let us illustrate the intuition of a physical judgment with an example. Consider the claim that a stone falling from a tower touches the ground after a time exactly determined by Galileo's formula. | What is meant by the concept "stone" 49
can in the end only be determined by empirical intuition; one has to look at a real stone or reproduce in the mind the perception created in this way to know to what the concept "stone" refers. Similar considerations apply to the process of falling and the other concepts that occur in the judgment. Further, the quantity of time can only be given by empirical intuition. Time as pure form of intuition has to be distinguished from any empirically perceived time variable, but here we are referring to a *particular* time variable. The meaning of the expression is the claim that the time of fall is equal to another, empirically observable time

empirisch wahrgenommenen Zeit behauptet wird, etwa der Zeit, die der Zeiger auf dem Zifferblatt einer Uhr braucht, um einen bestimmten Winkel zurückzulegen, oder der Zeit eines Bruchteils der Erdumdrehung. Diese *bestimmte* Zeitgröße kann ich immer nur durch unmittelbare Wahrnehmung erkennen.

Das mathematische Urteil dagegen betrifft Gegenstände, die durch reine Anschauung und reines Denken allein zu erfassen sind. Es ist das Charakteristikum solcher Gegenstände, daß sie restlos bestimmt sind, daß sie sich mit den Bestimmungsstücken des Denkens und der Anschauung vollständig erschöpfen lassen. Wenn ich von einer mathematischen Kugel spreche, so ist damit der Gegenstand lückenlos bestimmt, und seine sämtlichen Eigenschaften lassen sich aus seiner Vorstellung ableiten; wenn ich aber von dem Erdball spreche, so weiß ich zunächst noch gar nicht, welche Raumform damit gemeint ist. Zwar weiß ich, daß sie einer Kugel ähnlich sein wird, aber die leicht abgeplattete, viel gezahnte wirkliche Gestalt ist mir noch nicht gegeben; ich weiß zwar, daß der Erdball in einem bestimmten Zeitpunkt eine ganz bestimmte räumliche Gestalt hat, aber diese zu bestimmen, ist erst Aufgabe der Forschung, nicht durch die Vorstellung Erdball ohne weiteres gegeben. Dem empirischen Gegenstand haftet das eigentümlich irrationale Moment aller Wirklichkeit stets an.

In den mathematischen Urteilen kann deshalb der Gegenstand, eben weil er vollständig gedacht werden kann, zuvor *gesetzt* werden. Es wird nie eine Gültigkeit *schlechthin* behauptet, sondern die Aussage stets an das Vorhandensein bestimmter Bedingungen geknüpft; diese Bedingungen bestimmen restlos den
50 | Gegenstand, und darum läßt sich die Aussage mit dem Anspruch der *Gewißheit* daran knüpfen. Das physikalische Urteil aber kann nicht an *gesetzte* Bedingungen angeknüpft werden, sondern nur an Voraussetzungen, die letzten Endes von der Natur gegeben werden; sie lassen sich nicht vollständig übersehen und bestimmen, sondern müssen schlechtweg hingenommen werden. Z.B. von diesem Erdball, den ich eben nur wahrnehmen kann, nicht aus reiner Anschauung apriori konstruieren, wird etwas ausgesagt. Die Urteilsformen des hypothetischen und des kategorischen Urteils entsprechen diesen beiden Kategorien. Zwar ist es möglich, auch physikalischen Urteilen die sprachliche Form des hypothetischen zu geben, und umgekehrt, mathematische Urteile als Aussagen über einmalige, ganz bestimmte Gegenstände zu formulieren. Aber wenn Logik mehr sein soll als Grammatik, wird man den Begriff des hypothetischen Urteils auf solche Urteile beschränken, deren Bedingungen sich erschöpfend und eindeutig bestimmen lassen, und den Namen des kategorischen Urteils nur solchen Urteilen zukommen lassen, deren Geltungsbedingungen ihrer Natur nach nicht erschöpfend bestimmbar sind. Dann werden die mathematischen Urteile hypothetisch, die physikalischen Urteile kategorisch sein. Um jede Zweideutigkeit des Sprachgebrauchs zu vermeiden, wollen wir jedoch die ersteren als *Setzungs*urteile, die andern als *Wirklichkeits*urteile bezeichnen.

Trotz ihres empirischen Charakters enthalten aber die Wirklichkeitsurteile ein apriorisches Moment. Denn die *Relation*, die in ihnen behauptet wird und die ihnen überhaupt erst den Charakter des Urteils verleiht, ist nicht in der Wahrnehmung gegeben, sondern erst vom Denken hinzugefügt. Die begriffli-

variable, for example the time in which an arm on a clock covers a certain angle, or the time of a fraction of the revolution of the Earth. This *particular* time variable I can only come to know by direct observation.

In contrast, the mathematical judgment concerns objects that can be grasped by pure intuition and pure thought alone. It is a characteristic of such objects that they are completely specified by thought and intuition. When I speak of a mathematical sphere, then the object is completely determined and all its properties can be derived from the imagination; however, when I speak of the Earth, then initially I am unclear which geometrical shape is meant. I know that it is going to be similar to a sphere, but the slightly flattened real form with many bumps is not given; I know that at any fixed point in time the Earth has a particular shape, but to determine it is a matter of research and not given simply by the imagination of the Earth. The peculiar irrational aspect of reality always attaches to the empirical object.

In mathematical judgments the object can be *specified* initially because it can be fully grasped. There is never a claim of validity *par excellence*, rather the claim is always conditional on the presence of certain requirements; these requirements completely determine the | object and therefore the claim to *cer-* 50
*tainty* is secured. However, physical judgments cannot be based on *defined* conditions, but can only be based on preconditions that are ultimately given by nature; they cannot be fully grasped and determined, but must simply be assumed. For example, a claim is made about the Earth, which I can only perceive and not reconstruct a priori from pure intuition. The two forms of judgment, the hypothetical and the categorical, correspond to these two types. It is possible to give physical judgments the linguistic form of hypothetical judgments, and vice versa, to formulate mathematical judgments as assertions about particular unique objects. But if logic is going to be more than just grammar, one will restrict the term "hypothetical judgment" to such judgments whose preconditions can be determined exhaustively and uniquely and the name "categorical judgment" will only be assigned to such judgments whose conditions of validity cannot be completely determined due to the nature of the judgment. Thus, mathematical judgments are hypothetical and physical judgments are categorical. To avoid all ambiguity in language use we will call the first judgments *of definition* and the second judgments *of reality*.

Despite their empirical character, judgments of reality contain an a priori aspect because the *relation* they assert, and which gives them the form of a judgment, is not given in perception but is added by thought. The conceptual

che Erfassung, die aus der Wahrnehmung die Erkenntnis schafft, kommt erst durch den Akt der Synthesis zustande, der selbst nicht empirisch gegeben, sondern eine freie Setzung ist. Es ist das, was Kant Stellung unter die transzendentale Einheit der Apperzeption nennt. Aber Kant zeigt auch, daß dieser Akt nicht auf die Urteilsrelation beschränkt bleibt, sondern daß es derselbe Akt ist, der die Synthesis im Begriffe vollzieht und so den Gegenstand allererst konstituiert. So ist das physikalische Urteil nicht bloß eine Relation zwischen zwei schlechthin empirischen Begriffen, sondern diese selbst bedeuten bereits eine Einheitssetzung, eine Synthesis von Relationen.

51 Durch Begriffe des Verstandes allein aber ist der Gegenstand auch noch nicht bestimmt. Sondern erst seine Einordnung in die Form der Anschauung bestimmt ihn völlig. So liegt in dem empirischen Begriff Stein nicht nur die transzendentale Synthesis enthalten, sondern der Gegenstand wird zugleich als räumlich ausgedehnt vorgestellt. Zwar bedarf auch die Anschauungsform, insofern sie nach einem Ausdruck Kants auch „Gegenstand der reinen Anschauung" ist, der Synthesis durch die transzendentale Apperzeption, und ist insofern diesem „höchsten Punkt, an dem man allen Verstandesgebrauch, selbst die ganze Logik, und, nach ihr, die Transzendentalphilosophie heften muß"[16] unterstellt, dennoch aber ist jene Synthesis nicht hinreichend, um den Gegenstand aus sich zu konstituieren. Erst durch ihre Anwendung auf Wahrnehmungsinhalte in Raum und Zeit als Formen des Nebeneinander wird die Konstitution des Gegenstandes vollzogen. „Zum Erkenntnisse gehören zwei Stücke: erstlich der Begriff, dadurch überhaupt ein Gegenstand gedacht wird (die Kategorie), und zweitens die Anschauung, dadurch er gegeben wird; denn könnte dem Begriffe eine korrespondierende Anschauung gar nicht gegeben werden, so wäre er ein Gedanke der Form nach, aber ohne allen Gegenstand, und durch ihn gar keine Erkenntnis von irgendeinem Dinge möglich".[17] Der physikalische Vorgang wird nach einem Schema vorgestellt, das aus Raum und Zeit als reinen Anschauungsformen gebildet ist. So wird in dem benutzten Beispiel der Stein von zwar beliebiger, aber bestimmter räumlicher Form gedacht, seine Bahn als gerade Linie angesehen, seine Geschwindigkeit als gleichmäßig und stetig wachsend; es wird eine ideale Struktur in das wirkliche Geschehen hineingedacht. Von dem idealen Körper, der in dem idealen leeren Raum im ideal homogenen Kraftfeld fällt, ließe sich die Gültigkeit des Galileischen Gesetzes mit Gewißheit behaupten. In diesem Falle waren alle notwendigen Bedingungen erschöpfend und eindeutig *gesetzt*; aus ihnen läßt sich dann das Resultat mit mathematischen Methoden ableiten, und die Behauptung wäre ein zwar synthetischer, aber mathematischer Satz geworden. Das wesentlich Unterscheidende besteht für den empirischen Satz darin, daß in ihm *die Geltung einer bestimmten mathematischen Struktur*
52 *von der* | *letzten Grundes nur schlechthin gegebenen Wirklichkeit behauptet wird.*

Geltung scheint mir dasjenige Wort zu sein, das diesen eigentümlichen Sachverhalt am besten charakterisiert. Es wäre falsch, von Übereinstimmung

16. Kritik der reinen Vernunft. S. 109 (S. 134) Anm.
17. Kritik der reinen Vernunft. S. 116 (S. 146).

representation that creates knowledge from perception is only accomplished by the act of synthesis, which is not given empirically, but is a free definition. This is what Kant calls placing under the transcendental unity of apperception. But Kant also shows that this act is not restricted to the relation given by a judgment; but the same act accomplishes the synthesis of the concept of the object. Consequently, a physical judgment is not just a relation between two purely empirical concepts; rather these already represent a determination of units, a synthesis of relations.

However, the object is not yet fully determined by concepts given by the un- 51
derstanding. Only its integration into the form of intuition determines it completely. For example the empirical concept "stone" not only contains the transcendental synthesis, but the object is also thought of as extended in space. Although the form of intuition requires – in so far as it is in Kant's terms also "an object of pure intuition" – the synthesis by transcendental apperception and is thereby subordinate to this "highest point on which all use of the understanding, even all of logic and, following it, transcendental philosophy, depends",[16] the synthesis is not sufficient to constitute the object. The constitution of the object is only completed by the application of the content of perception in time and space as forms of extension. "There are two parts to knowledge: first, the concept, so that it is possible to think of some object (the category), and second, the intuition, by which the concept is given; since if one could not give an intuition corresponding to the concept, then it would have the form of a thought without any actual object, and therefore would not permit any knowledge of things".[17] The physical process is presented in a schema that is formed from space and time as pure forms of intuition. Hence, in the above example, the stone is considered to have arbitrary but definite spatial shape, its trajectory is seen as a straight line, its velocity as increasing uniformly and continuously: an ideal structure is imagined in the real process. The validity of Galileo's law could be asserted with certainty about the ideal object that is falling in an ideal empty space in an ideal homogeneous force field. In this case, all necessary conditions would be completely and uniquely *specified*; the result can then be derived by mathematical methods, and the claim would have become indeed a synthetic, but mathematical, assertion. The essential difference for the empirical claim lies in the fact that it *asserts the validity of a particular mathematical structure of | reality.* 52

"Validity" seems to me to be the word that best characterizes this situation. It would be incorrect to speak of a correspondence between the mathematical

16. *Critique of Pure Reason*. p. 109 (p. 134), footnote.
17. *Critique of Pure Reason*. p. 116 (p. 146).

der mathematischen Struktur mit der Wirklichkeit zu reden; diese kann nie behauptet werden, da das wirkliche Geschehen stets durch unendlich viele Gleichungen bestimmt ist. Es gibt, wie schon früher ausgeführt wurde, keine abgeschlossenen Vorgänge in der Natur. So wird das mathematische System stets nur mit Näherung die Wirklichkeit darstellen. Aber es muß behauptet werden, daß die das Geschehen wesentlich bestimmenden Einflüsse in den Gleichungen dargestellt werden. Das „wesentlich" läßt sich hier direkt gleich „zahlenmäßig überwiegend" setzen, so daß behauptet wird, durch die bestimmten mathematischen Gleichungen sei die zahlenmäßig meßbare Größe der auftretenden Objekte bis auf geringe Fehler bestimmt. Wenn man diese Annahme nicht machen will, ist empirische Erkenntnis überhaupt unmöglich. Dann wäre es uns versagt, irgendwie sinnvolle Aussagen über die Wirklichkeit zu machen; es ließe sich keinerlei Beziehung zwischen den mathematischen Gleichungen, die Relationen zwischen Gegenständen der reinen Anschauung bedeuten, und der Wirklichkeit herstellen. Wenn es physikalische Urteile geben soll, so muß eine näherungsweise Darstellung der Wirklichkeit durch mathematische Gleichungen möglich sein. Wir wollen in jedem Einzelfall eines solchen Sachverhalts von *Geltung* bestimmter mathematischer Gleichungen für die Wirklichkeit sprechen.

In der Tat stellen die Theorien der modernen Physik solche Gleichungssysteme dar, deren Geltung von der Wirklichkeit behauptet wird. Die Maxwellsche Theorie der Elektrizität stellt vier Grundgleichungen an die Spitze und leitet dann sämtliche Erscheinungen der Elektrizitätslehre als deren Folge ab. Wenn diese Gleichungen das wirkliche Verhalten mit Genauigkeit darstellen würden, so würden auch sämtliche übrigen Gesetze der Elektrizität mit Notwendigkeit zutreffen. Aber das kann nicht behauptet werden; für alles wirkliche Geschehen stellt das gesamte Gleichungssystem der Theorie nur eine mathematische Struktur dar, die die Wirklichkeit nicht restlos erschöpft. Zum Naturgesetz wird die Theorie dadurch, daß sie den Anspruch der *Geltung von der Wirklichkeit* er-
53 heben darf, d. h. daß sie das wirkliche | Verhalten mit zahlenmäßiger Näherung darstellt. Gerade die Maxwellsche Theorie ist ein Beispiel dafür, wie ein ausgedehntes mathematisches System weite Gebiete der Wirklichkeit einheitlich darstellen kann, ohne sie vollkommen zu erschöpfen; ihre Bedeutung besteht darin, daß sie die mathematische Struktur zwischen Gebieten aufdeckt, die früher ganz getrennt nebeneinander bestehen mußten.

Die Möglichkeit der physikalischen Erkenntnis ist damit auf die Angebbarkeit zahlenmäßiger Näherungswerte zurückgeführt. Dagegen läßt sich der Einwand erheben, daß die Bestimmung von Zahlen gar nicht Aufgabe der Physik sei, daß sie vielmehr nur die allgemeinen Gesetze aufzustellen habe; die Beschreibung der besonderen Wirklichkeit sei Aufgabe einer rein beschreibenden Disziplin, die im Rickertschen Sinne eine „Geschichte" der Wirklichkeit zu entwickeln habe. Man verdeutlicht diesen Unterschied auch gern an einem Beispiele. So sei es Aufgabe der Physik, das allgemeine Gasgesetz $pv = RT$ aufzustellen; aber die Werte von $p, v$ und $T$ im besonderen Fall zu messen, sei höchstens methodisches Hilfsmittel der Physik, nicht mehr ihre Aufgabe. Dieser Einwand verkennt jedoch den Zusammenhang von Einzelfall und allgemeinem Gesetz.

structure and reality; this can never be claimed, since real events are always governed by an infinity of equations. There are, as has been mentioned earlier, no closed processes in nature. Hence, the mathematical system will only ever approximately represent reality. But it has to be said that the influences that crucially determine the events are represented in the equations. Here "crucial" can be set equal to "numerically dominating," so we assert that the mathematical equations determine the numerically measurable quantity of the real objects down to a small error. If one is unwilling to make this assumption, then empirical knowledge is impossible in general. Then we would be unable to make any meaningful claims about reality; no relation could be established between reality and the mathematical equations that represent relations between objects of pure intuition. If we are to have physical judgments, then an approximate representation of reality by mathematical equations has to be possible. In each such circumstance, we will speak of the *validity* of particular mathematical equations for reality.

In fact, the theories of modern physics present such systems of equations, which are claimed to be valid in reality. Maxwell's theory of electricity has four equations as its basis and then derives all phenomena in the theory of electricity as their consequences. If these equations accurately represented real events, then all derivative laws of electricity would necessarily hold. But this cannot be claimed; the system of equations of the theory only represents a mathematical structure for the real events that does not fully account for reality. The theory becomes natural law when its claim to *validity in reality* can be asserted, i. e.,
that it represents the real | processes to numerical approximation. Maxwell's 53
theory is a particularly good example of how an extensive mathematical system can give a unified account of large areas of reality, without fully exhausting it; the importance of the theory consists in the fact that it reveals the mathematical structure between areas that were previously considered distinct.

The possibility of physical knowledge has thereby been traced back to the assertability of numerical approximations. The objection can be raised that the duty of physics is the postulation of general laws, not the determination of numbers; the description of reality is the task of a purely descriptive discipline, that has to develop – to use Rickert's term – a "history" of reality. This difference is often illustrated by an example. So it would be the task of physics to postulate the general law of gases $pv = RT$, but to determine the values of $p, v$ and $T$ in any particular case would be at most a methodological aid to physics, not its task. However, this objection mistakes the relationship between the individual case

Auch das Boylesche Gesetz gilt nicht für alle Gegenstände schlechthin, sondern nur für eine Klasse von Gegenständen, für Gase; ja nicht einmal für alle Gase, sondern nur für eine Klasse von Gasen, nämlich für stark verdünnte. Trotzdem wird für diese Klasse die ihr eigentümliche Konstante $R$ zahlenmäßig bestimmt. Für andere Gase wäre $R$ nicht mehr als Konstante anzusehen, sondern würde eine komplizierte Funktion werden, die erst wieder neu bestimmt werden müßte.[18] Es werden also zahlenmäßige Bestimmungen getroffen, die nicht für die ganze Natur, sondern für eine enge Klasse von Gegenständen gelten; diese Ge-
54 genstände sind letzten Endes auch nur aufzeigbar, nur | durch empirische Anschauung zu erkennen. Der Spezialisierung aber ist nirgends eine Grenze gesetzt. Immer engere Klassen von Gegenständen wird die Physik gesetzmäßig zu beherrschen suchen müssen; immer mehr unbestimmte Größen, die als Buchstaben in den Formeln auftreten, wird die Physik zahlenmäßig feststellen und bestimmten Gegenständen der empirischen Anschauung zuordnen müssen. Der Einzelfall wird zwar so nie erreicht, aber die Richtung auf ihn wird stets innegehalten; und die Messung der sogenannten Naturkonstanten bedeutet nichts anderes als ein Fortschreiten in der Richtung, an deren Endpunkt der Einzelfall steht.

Die umgekehrte Forschungsrichtung wird ebenso von der Physik innegehalten. Jede Konstante wird wieder als Funktion hingestellt; die für bestimmte Gesetze als rein gegeben auftretende Naturkonstante, deren Messung zahlreiche Experimente gewidmet werden, wird mit ganz anderen Größen in Zusammenhang gebracht, so daß sie als Funktion erscheint, deren spezieller Wert in den ersteren Gesetzen nur durch besondere Umstände zustande kommt. So ist die Galileische Konstante des Fallgesetzes, $g = 9,81$, solch eine schlechthin gegebene Naturkonstante. Die große Leistung Newtons ist es, sie in Beziehung gebracht zu haben zur Umlaufszeit des Mondes um die Erde und damit zur Planetenbewegung überhaupt. Die frühere Konstante erscheint jetzt als Funktion der Entfernung vom Erdmittelpunkt. Das ist allgemein der Weg der Physik: Konstanten in Funktionen aufzulösen, allgemeinere Gesetze zu finden, als deren bloßer Spezialfall das frühere Gesetz erscheint.[19] Ein Ende dieses Prozesses ist nicht abzusehen.

Es hätte aber keinen Sinn, eine der beiden Forschungsrichtungen zu bevorzugen, etwa den Weg zum allgemeineren Gesetz allein physikalische Erkenntnis zu nennen. Denn dann wäre der Bestand an physikalischen Sätzen abhängig von

18. Diese Darstellung der Abänderung des Boyleschen Gesetzes weicht allerdings von der üblichen ab, ihre Rechtmäßigkeit ist jedoch leicht zu erkennen. Gewöhnlich wird die Form der Gleichung durch Einführung von Zusatzgliedern geändert, so daß die Größe $R$ als Konstante erhalten bleibt. Doch ließe sich die neue Gleichung mit beliebiger Näherung auch so schreiben, daß sie die Form

$$pv = F(p, v, T, a_1, a_2, \dots)T$$

annimmt, so daß $F$ für bestimmte Werte der neuen Konstanten $a_1, a_2, \dots$ den alten Wert $R$ erhält.

19. Vgl. hierzu Cassirer, S. 351 ff.

and the general law. Boyle's law does not apply to all objects in general, but only to a class of objects, to gases; indeed, not even to all gases, but only to the class of strongly diluted gases. Nonetheless, one determines numerically the constant $R$ particular to this class. For other gases, $R$ would not be a constant, but would be a complicated function that itself would have to be determined anew.[18] Hence, there are numerical determinations that do not apply to all of nature, but only to a narrow class of objects; ultimately only these objects can be detected, only | 54
these can be known through empirical intuition. But there is no limit to the specialization. Physics will have to aim to capture in laws ever more narrow classes of objects and physics will have to assign to particular objects of empirical intuition increasing numbers of unmeasured variables appearing as letters in formulas, whose values must be determined numerically. The individual case will never be reached in this way, but the progression towards it will be maintained; the measurement of a so-called natural constant does not mean anything other than progress along the path whose endpoint is the individual case.

The opposite direction of research is also followed by physics. Every constant is presented as a function; the natural constant which is simply given for certain laws and to whose measurement several experiments are dedicated is brought into connection with completely different quantities, so that it appears as a function whose specific value in the previous laws is only attained under special circumstances. For example Galileo's constant in the law of falling bodies, $g = 9.81$, is such a plainly given natural constant. Newton's great achievement was to have shown its connection with the orbital period of the moon around the Earth and thereby with planetary motion in general. The earlier constant appears now as a function of the distance to the center of the Earth. This is the general approach of physics: to resolve constants into functions, to find more general laws that contain the previous law as a special case.[19] No end of this process is in sight.

It would make no sense to favor any of these two directions of research; for example to only deem the path to more general laws physical science, since then the stock of physical theorems would depend on the arbitrary starting point of

18. This presentation of the variant of Boyle's law diverges from the common one, but its correctness can easily be seen. Normally the form of the equation is changed by the introduction of additional variables, so that the quantity $R$ remains a constant. But the new equation could also be written with arbitrary approximation in the form

$$pv = F(p, v, T, a_1, a_2, \ldots)T$$

so that $F$ assumes the old value $R$ for particular values of the new constants $a_1, a_2, \ldots$.

19. See Cassirer, p. 351 ff.

dem zufälligen Anfangspunkt der Forschung; speziellere Sätze als die, mit denen begonnen wurde, wären keine physikalischen Sätze. Eine derartige Willkür aber darf keinesfalls über den Inhalt einer Wissenschaft entscheiden. Es muß vielmehr daran festgehalten werden, daß physikalische Urteile sich auf Gegenstandsklassen von jeder beliebigen Enge oder Weite erstrecken. Klassen bleiben
55 | es stets; das Ganze des Universums in eine physikalische Gleichung zu fassen, ist ebenso unmöglich wie den Einzelfall restlos in mathematische Beziehungen aufzulösen.[20] *Die Zuordnung einer jeden, zunächst nur als System von mathematischen Sätzen gegebenen Klasse zu Gegenständen der empirischen Anschauung ist die eigentliche Aufgabe der Physik; sie schließt die zahlenmäßige Bestimmung der Konstanten ein.*

Auf den Begriff der *Möglichkeit* muß an dieser Stelle noch eingegangen werden. Es soll ein Vorgang unmöglich genannt werden, dessen Vorkommen in der Erfahrung mit Gewißheit ausgeschlossen ist. Danach sind zunächst alle Vorgänge unmöglich, die den Gesetzen der Logik widersprechen; aber auch diejenigen, die zu den synthetischen Grundsätzen der reinen Anschauung und des Denkens im Widerspruch stehen. Dies letztere schließt alle Vorgänge von der Existenz aus, die mathematischen Gesetzen widersprechen. Es wird nun vielfach noch eine dritte Forderung aufgestellt, die ein Vorgang erfüllen muß, um in der Erfahrung auftreten zu können: er darf den empirischen Gesetzen der Naturwissenschaft nicht widersprechen. Erst wenn er diese Bedingung erfüllt, so führt z. B. v. Kries aus, ist der Vorgang „objektiv möglich". Diese Behauptung impliziert den Satz, daß Vorgänge denkbar seien, die den mathematischen Gesetzen nicht widersprechen, aber dennoch zu empirischen Gesetzen im Widerspruch stehen; oder umgekehrt gefaßt, daß Vorgänge zu empirischen Gesetzen im Widerspruch stehen könnten, ohne zu mathematischen Gesetzen im Widerspruch zu sein.

Nach der im früheren gegebenen Analyse des physikalischen Urteils ist der Fehler dieser Ansicht leicht einzusehen. Ein physikalischer Satz behauptet von zwei Gegenstandskomplexen je eine bestimmte mathematische Struktur, und die Relation dieser Komplexe ist dann eine lediglich mathematische Notwendigkeit. Ein Widerspruch zu einem solchen Satz würde also nur dadurch zustande kommen, daß die mathematischen Strukturen zweier Gegenstandskomplexe nicht in einen mathematischen Zusammenhang zu bringen sind. Ein solcher
56 Vorgang wäre allerdings unmöglich, | aber auch gerade nur deshalb, weil mathematische Gesetze ihn ausschließen. Die Existenzunmöglichkeit eines Vorgangs läßt sich prinzipiell darauf zurückführen, daß der Vorgang zu mathematischen Gesetzen in Widerspruch steht. Der Satz, den die obige Annahme impliziert, ist also falsch, und die Annahme einer besonderen „physischen" Unmöglichkeit, die erst tatsächlich das Vorkommen in der Erfahrung ausschließt, unrichtig. Die „metaphysische" Möglichkeit ist nicht nur notwendige, sondern auch hinreichende Bedingung dafür, daß ein Vorgang von dem Vorkommen in der Erfahrung nicht ausgeschlossen ist.

20. Es ist übrigens leicht einzusehen, daß die Lösung des einen Problems die des andern in sich schlösse.

research. More specific theorems than those with which we started would not be theorems of physics. But such arbitrariness should never decide the content of a science. Rather, one has to maintain that physical judgments refer to classes of objects of any breadth or restriction. They always remain classes; | it is just as 55
impossible to subsume the whole universe under one physical equation as it is to fully capture the individual case in mathematical relationships.[20] *The actual task of physics is the coordination of a class, which is initially only given as a system of mathematical theorems, with objects of empirical intuition; this includes the numerical determination of constants.*

We still have to take a more detailed look at the concept of *possibility*. A process is called impossible if its occurrence in experience can be excluded with certainty. Accordingly, all processes are impossible that contradict the laws of logic, as are those that contradict the fundamental synthetic principles of pure intuition and thought. This latter point excludes from existence all processes that contradict mathematical principles. Often a third requirement is set for a process to occur in experience: it may not contradict the empirical laws of natural science. For example, von Kries explains that only when this condition has been satisfied is the process "objectively possible". This assertion entails the claim that there are thinkable processes that do not contradict the laws of mathematics but do contradict empirical laws; or conversely, that there are conceivable processes that can conflict with empirical laws without contradicting mathematical laws.

In light of the analysis of the physical judgment given previously the error of such a view is obvious. A physical claim postulates a mathematical structure for each of two complexes of objects and the relation of these complexes is then a mathematical necessity. A contradiction to such a claim could only arise if the mathematical structures of two complexes of objects could not be brought into connection. However, such a process would be impossible | precisely because the 56
mathematical laws preclude it. The impossibility of the existence of a process can in general be traced back to a contradiction between the process and mathematical laws. Hence, the claim that the above assumption makes is false and the assumption of a special "physical" possibility that rules out the occurrence of a process in experience is incorrect. The "metaphysical" possibility is not only a necessary but also a sufficient condition that the occurrence of a process is not ruled out in experience.

20. It can easily be seen that the solution of the one problem would include the solution to the other.

Die hier gegebene Definition des Möglichkeitsbegriffes stimmt mit der Kantschen überein. Kant definiert: „Was mit den formalen Bedingungen der Erfahrung (der Anschauung und den Begriffen nach) übereinkommt, ist *möglich*".[21]

Kries zieht zur Rechtfertigung seines Möglichkeitsbegriffes die Unterscheidung von „nomologischen" und „ontologischen" Bestimmungsstücken heran. Zur mathematischen Bestimmung eines Vorgangs gehört zweierlei: erstens die Differentialgleichungen und zweitens die Konstanten, die durch die Anfangsbedingungen des Vorgangs gegeben sind. Die ersteren enthalten die allgemeine Form des Gesetzes, die letzteren die Zahlenwerte des besonderen Falles; zur Charakterisierung dieses Unterschiedes wählt Kries die obigen Bezeichnungen. Er meint, ein Vorgang sei dann möglich, wenn er durch keine nomologische Bestimmung ausgeschlossen ist, d. h. durch keine Differentialgleichung der Physik; ob er wirklich ist, hängt aber davon ab, ob seine ontologischen Bestimmungsstücke mit dem allgemeinen Wirklichkeitszusammenhang vereinbar sind. Diesen Begriff der Möglichkeit, der durch Übereinstimmung mit nomologischen Bedingungen definiert ist, nennt er physische oder objektive Möglichkeit. Er übersieht nur, daß dieser Begriff der Möglichkeit auch nichts anderes bedeutet als Übereinstimmung mit mathematischen Gesetzen. Die Differentialgleichungen der Physik sind eben auch nur mathematische Beziehungen. Nicht darin besteht die Entdeckung von Naturgesetzen, daß unter den mathematischen Gleichungen solche ausgesucht werden, die auf die Natur angewandt werden können. Es muß vielmehr prinzipiell vorausgesetzt werden, daß zu jeder
57 mathe|matischen Gleichung auch ein Vorgang möglich ist, der durch sie dargestellt wird. Das ist ein Grundsatz, der in der Physik häufig angewandt wird; so werden in der Hydrodynamik auf rein mathematischem Wege Lösungen der Lagrangeschen Grundgleichungen gesucht, und erst nachher wird festgestellt, welchen wirklichen Vorgang man eigentlich darin beschrieben hat. Daß die Mathematik überhaupt auf die wirklichen Gegenstände anwendbar ist, ist selbst kein empirischer Satz, sondern methodische Voraussetzung der Physik; sie bedeutet nichts anderes als den großen Grundgedanken der Kantschen Erkenntnistheorie. *Erst die Zuordnung bestimmter mathematischer Gleichungen zu bestimmten, nur durch empirische Anschauung aufzuzeigenden Gegenständen ist der Inhalt der physikalischen Erkenntnis.* Kries' Möglichkeitsbegriff ist kein anderer als der der metaphysischen Möglichkeit. Sein Wirklichkeitsbegriff aber ist sehr unklar: ein Vorgang kann auch in seinen ontologischen Bestimmungsstücken so beschaffen sein, daß sie den gegebenen Zuständen des Wirklichkeitsablaufs nicht widersprechen, und braucht deshalb immer noch nicht wirklich zu sein. Wirklich ist er erst, wenn er selbst oder ein nach mathematischen Gesetzen mit ihm verknüpfter Vorgang empirisch wahrgenommen wird. Ein Vorgang, der nur nomologisch bestimmt ist, ist eben noch unbestimmt, kann noch gar nicht Vorgang genannt werden; denn möglich sind immer nur vollkommen bestimmte Vorgänge. Eine derartige Bezeichnung eines Vorgangs durch bloße no-

21. Kritik der reinen Vernunft, S. 185 (S. 265).

The definition of the concept of possibility given here agrees with Kant's. Kant defines: "Whatever agrees with the formal conditions of experience (in terms of intuition and concepts), is *possible*".[21]

Kries justifies his concept of possibility by the distinction between "nomological" and "ontological" aspects of science. There are two parts to the mathematical determination of a process: first, the differential equations and, second, the constants that are given by the initial conditions of the process. The former contain the general form of the law, the latter the numerical values of the particular case. Kries chose the above terms to characterize this distinction. He thinks that a process is possible if it is not ruled out by a nomological condition, i. e., by no differential equation of physics; whether it is real depends on whether its ontological features are compatible with the general context of reality. He calls this type of possibility, which is defined by its compatibility with nomological conditions, physical or objective possibility. What he does not realize is that this type of possibility means nothing other than compatibility with mathematical laws. The differential equations of physics are simply mathematical relations, too. The discovery of the laws of nature does not consist in selecting among mathematical equations those that can be applied in nature. Rather it has to be assumed in principle that for every mathematical | equation there 57
is a process that can be represented by it. This is a principle that is often applied in physics; for example in hydrodynamics, solutions of the fundamental Lagrangian equations are sought on a purely mathematical basis, and only later does one realize which real process was described by them. It is a methodological assumption of physics and not an empirical claim that mathematics is applicable to real objects at all, which means the same thing as the basic thought of Kant's theory of knowledge: *The content of physical knowledge is the coordination of particular mathematical equations to particular objects present in empirical intuition.* Kries's concept of possibility is no other than that of metaphysical possibility. But his concept of reality is very unclear: a process may have ontological features that do not contradict the given unfolding of reality, but it still need not be real. It is only real if it, or a process connected to it by mathematical laws, is empirically perceived. A process that is only nomologically determined is still indeterminate and therefore cannot yet be called a process because only fully determined processes are possible. Such a description of a process by

21. Critique of Pure Reason, p. 185 (p. 265).

mologische Bestimmungsstücke bedeutet immer nur die generalisierende Zusammenfassung in einer Klasse. Dies ist die eigentliche Bedeutung jener Unterscheidung. Die Klasse ist durch die Form der Differentialgleichungen und die in ihnen auftretenden Konstanten, die bereits zahlenmäßig bestimmt sein müssen, gegeben; die Integrationskonstanten werden noch unbestimmt gelassen, und ihre besonderen Werte entsprechen je einem Exemplar der Klasse. Aber ein Exemplar ist nicht deshalb wirklich, weil seine Konstanten besondere Werte angenommen haben; und es ist auch nicht deshalb bloß möglich, weil diese Werte noch offen gelassen sind. Unbestimmtheit ist nicht das Kriterium der Möglichkeit, und Bestimmtheit nicht die Definition der Wirklichkeit. Sondern die Bedeutung dieser fundamentalen Kategorien läßt sich letzten Grundes nur als offene Einsicht, als reines Schauen erfassen.

58 Im Zusammenhang mit der Unklarheit des Möglichkeitsbegriffs steht der vielfach verbreitete Irrtum, daß man das Universum als die eine wirkliche Welt unter zahlreichen möglichen Welten betrachtet. Man geht dabei von der Unterscheidung in Differentialgleichungen und Anfangsbedingungen aus und bildet so den Begriff einer Klasse von Welten, in der die eine wirkliche Welt die Stellung eines Exemplares einnimmt. Der Fehler dieser Auffassung liegt darin, daß man auf unendliche Systeme überträgt, was nur für endliche Systeme Gültigkeit hat. Für endliche Systeme gilt es allerdings, daß man jeden wirklichen Vorgang als Exemplar einer Klasse betrachten kann, und daß dann die ganze Klasse mögliche Vorgänge darstellt. Diese Verallgemeinerung ist aber nur deshalb möglich, weil jedes endliche System durch eine endliche Zahl von Gleichungen wesentlich (d. h. zahlenmäßig überwiegend) bestimmt ist. Sie wird falsch, wenn wie bei dem Universum dies nicht mehr erfüllt ist. Darum hat es keinen Sinn, den Begriff einer Klasse zu bilden, in der das Weltall ein Exemplar ist; und die Annahme mehrerer möglichen Welten ist deshalb unstatthaft.

Aber selbst wenn man einmal mit dem Begriff einer Klasse von Welten arbeiten wollte, würde das Fehlerhafte der Betrachtungsweise noch an einer eigentümlichen Konsequenz hervortreten. Es ist nach dem Vorhergehenden einleuchtend, daß wenn eine Klasse von Vorgängen möglich ist, auch mehr als ein Exemplar wirklich sein kann. Denn der Gegensatz „Klasse und Exemplar", der identisch ist mit dem Gegensatz „Differentialgleichungen und Anfangsbedingungen", fällt keineswegs zusammen mit dem Gegensatz „möglich und wirklich". Es ist sogar leicht einzurichten, daß eine ganze Klasse von Vorgängen wirklich wird, etwa wenn man ein Gas sämtliche Kompressionszustände kontinuierlich durchlaufen läßt. Im Gegenteil, wenn nur ein einzelnes Exemplar wirklich wird, so bedarf dies noch einer besonderen Begründung aus den umgebenden Umständen. So mußte die oben genannte Auffassung es noch besonders begründen, warum unter der Klasse möglicher Welten gerade nur *eine* wirklich geworden ist; es hätte ja auch sein können, daß zwei oder mehr Welten wirklich geworden wären. Das Absurde einer solchen Forderung leuchtet sofort ein; denn die Gesamtheit aller Erfahrung ist eben notwendig eine Einheit. Der Feh-
59 ler rührt daher, daß man diese schlechthin einzige | Einheit der Erfahrungswelt zum Einzelfall einer Klasse gemacht hat. Die Möglichkeit einer Klasse von Fäl-

purely nomological features is always a generalized summary of a class. This is the actual meaning of the distinction. The class is given by the form of the differential equations and the constants that occur in them that already must have been numerically determined; the integration constants are left unspecified and their particular values correspond to one instance of the class. But an instance is not real because the constants have assumed particular values; and it is not possible just because the values have been left undetermined. Lack of determination is not the criterion for possibility, and determination is not the definition of reality. The meaning of these fundamental categories can ultimately only be grasped as plain insight, as pure intuition. In the context of the confusion sur- 58
rounding the concept of possibility there is the widely spread mistake that the universe is the one real world among many possible worlds. This uses the distinction between differential equations and initial conditions to construct a class of worlds of which the real world is but one instance. The mistake is that one is transferring to infinite systems what holds only for finite systems. For finite systems it is true that one can view every real process as an instance of a class and that the class represents all possible processes. But this generalization is only possible because any finite system is essentially (i. e., in numerical terms) determined by a finite number of equations. It becomes false if – as in the case of the universe – this is not satisfied. Hence, it makes no sense to create the concept of a class of which the universe is an instance; and consequently the assumption of several possible worlds is not permitted.

But even if one were to work with the concept of a class of worlds, a peculiar consequence would show the mistake in the approach. It is obvious from the previous discussion that if a class of processes is possible more than one instance of it may really occur. After all, the contrast between "class and instance" is identical to the contrast between "differential equation and initial condition", but it by no means corresponds to the contrast between "possible and real". It is easy to make a whole class of processes possible, for example if one lets a gas pass continuously through all states of compression. In contrast, if only one instance becomes real, then this needs special justification in the given circumstances. Consequently, the above mentioned view would have to explain why in the whole class of possible worlds just *one* became real; after all, it could have also happened that two or several worlds became real. The absurdity of such a demand is obvious since the entirety of experience is necessarily a unity. The error results from the fact that one has made this single | unity of experi 59
ence an instance of a class. The possibility of a class of instances consists in the

len besteht nur darin, daß jeder einzelne ein Glied in der allumfassenden Gesamtheit der Erfahrung sein kann; auf diese Gesamtheit ist der Begriff der Möglichkeit nicht anwendbar.[22]

Es ist im früheren ausgeführt worden, daß die physikalische Erkenntnis in der Zuordnung mathematischer Gleichungen zu bestimmten Gegenständen der empirischen Anschauung besteht. Es ist ferner gezeigt, daß für diese Zuordnung die Bestimmung einer Zahl, der die Klasse charakterisierenden Konstanten, erforderlich ist; und daß eine näherungsweise Bestimmbarkeit dieser Zahl vorausgesetzt werden muß. Auf diesen Begriff der Näherung muß jetzt weiter eingegangen werden.

Die Darstellung durch Gleichungen, so war entwickelt worden, kann ein wirkliches Geschehen nicht erschöpfen; sie kann immer nur eine zahlenmäßige Näherung erreichen. Es wird also behauptet, daß gewisse Einflüsse ein Geschehen wesentlich bestimmen, und daß die anderen, die man gerade nicht gemessen hat, dagegen verschwinden. So stellt man die Bewegung eines fallenden Steines dar, indem man Gravitation, Luftwiderstand, Fallhöhe in den Gleichungen berücksichtigt; die geringen Einwirkungen etwa in der Nähe vorhandener großer Massen oder elektromagnetische Einflüsse vernachlässigt man. *Über die*
60 *Größe von | Einwirkungen, deren Zahlenwert und Gesetz man nicht kennt, wird also eine Aussage gemacht.* Und das ist wesentlich für physikalische Urteile; ohne eine solche Aussage würden sie gar nicht zustande kommen. Nun dürfen zwar solche Behauptungen nicht beliebig aufgestellt werden. Sondern das ist gerade die Aufgabe der Erfahrung, festzustellen, welches die wesentlichen Einflüsse sind und welche anderen vernachlässigt werden dürfen. Diese Erfahrung kann natürlich nur durch Messungen gemacht werden; in ihnen wird festgestellt, ob die theoretisch berechneten Zahlen den in der Wirklichkeit auftretenden nahekommen. Man wende hier nicht ein, daß Messungen dem Einzelfall gelten und aus dem Rahmen der physikalischen Wissenschaft herausfallen; denn es war bereits nachgewiesen, daß die Messung einzelner, die Klasse charakterisierender Konstanten zur Wissenschaft notwendig gehört. Wenn aber im physikalischen Satz eine Konstante für eine Klasse von Fällen als gültig behauptet wird, so muß

22. Vgl. hierzu Kritik der reinen Vernunft. S. 196 (S. 283). „Sonst ist die Armseligkeit unserer gewöhnlichen Schlüsse, wodurch wir ein großes Reich der Möglichkeit herausbringen, davon alles Wirkliche (aller Gegenstand der Erfahrung) nur ein kleiner Teil sei, sehr in die Augen fallend. Alles Wirkliche ist möglich; hieraus folgt natürlicherweise, nach den logischen Regeln der Umkehrung, der bloß partikulare Satz: einiges Mögliche ist wirklich, welches denn so viel zu bedeuten scheint, als: es ist vieles möglich, was nicht wirklich ist. Zwar hat es den Anschein, als könne man auch geradezu die Zahl des Möglichen über die des Wirklichen dadurch hinaussetzen, weil zu jener noch etwas hinzukommen muß, um diese auszumachen. Allein dieses Hinzukommen zum Möglichen kenne ich nicht. Denn was über dasselbe noch zugesetzt werden sollte, wäre unmöglich. Es kann nur zu meinem Verstande etwas über die Zusammenstimmung mit den formalen Bedingungen der Erfahrung, nämlich die Verknüpfung mit irgendeiner Wahrnehmung hinzukommen; was aber mit dieser nach empirischen Gesetzen verknüpft ist, ist wirklich, ob es gleich unmittelbar nicht wahrgenommen wird. Daß aber im durchgängigen Zusammenhange mit dem, was mir in der Wahrnehmung gegeben ist, eine andere Reihe von Erscheinungen, mithin mehr als eine einzige alles befassende Erfahrung möglich sei, läßt sich aus dem, was gegeben ist, nicht schließen."

requirement that each instance can be one link in the all-embracing entirety of experience; the concept of possibility is not applicable to this entirety.[22]

As observed previously, physical knowledge consists in the coordination of mathematical equations with particular objects of empirical intuition. It has further been shown that this coordination requires the determination of a number, namely the constant characterizing the class; and one has to assume that an approximation to this number can be determined. We now have to consider this notion of approximation a little more closely.

We found that a representation by equations was insufficient to exhaust a real event; only a numerical approximation can be attained. Hence, the claim is that certain influences crucially determine an event, and that others that one has not measured disappear in comparison. For example the movement of a falling stone is described by including gravitation, air resistance and height in the equations; other lesser influences, such as the presence of large masses or electro-magnetic influences, are neglected. *Hence a claim is made about the size of | influences whose numerical values and laws one does not know.* This 60
is essential to physical judgments; without such a claim they would never come about. Of course, such claims may not be made arbitrarily. Rather, it is precisely the task of experience to establish which are the important influences and which others may be neglected. Naturally, this experience can only be obtained by measurements; they establish whether the theoretically calculated numbers are anywhere close to the numbers occurring in reality. One may not object that measurements apply to the instance and do not fall into the space of physical science, since it has already been shown that the measurement of individual constants characterizing the class necessarily forms part of science. If there is a physical principle that claims that a constant holds for some class of cases, then

22. See the *Critique of Pure Reason*, p. 196 (p. 283). "Moreover, the poverty of the customary inferences through which we throw open a great realm of possibility, of which all that is actual (the objects of experience) is only a small part, is patently obvious. Everything actual is possible; from this proposition there naturally follows, in accordance with the logical rules of conversion, the merely particular proposition, that some possible is actual; and this would seem to mean that much is possible which is not actual. It does indeed seem as if we were justified in extending the number of possible things beyond that of the actual, on the ground that something must be added to the possible to constitute the actual. But this [alleged] process of adding to the possible I refuse to allow. For that which would have to be added to the possible, over and above the possible, would be impossible. What can be added is only a relation to my understanding, namely that in addition to agreement with the formal conditions of experience there should be connection with some perception. But whatever is connected with perception in accordance with empirical laws is actual, even although it is not immediately perceived. That yet another series of appearances in thoroughgoing connection with that which is given in perception, and consequently that more than one all-embracing experience is possible, cannot be inferred from what is given." (Translation by Norman Kemp Smith, 1968)

diese Konstante auch für den Einzelfall gelten; und es muß daher auch für den bestimmten, einzelnen Vorgang behauptet werden, daß eine Messung der Konstanten für ihn zu einem zahlenmäßigen Näherungswert führt. Die physikalische Erfahrung besteht deshalb in der Konstatierung einer zahlenmäßigen Näherung durch das Experiment, und umgekehrt behauptet der physikalische Satz kraft der Empirie die annähernde Geltung bestimmter Zahlen für eine Klasse von Naturgeschehnissen.

Das ist das Eigentümliche der Erfahrung, daß sie eine Regel aufstellt für Wirkungen, die in ihrer Vielseitigkeit und Größe ganz unkontrollierbar sind. Alles andere am physikalischen Satz ist mathematische Relation; dies eine hinzukommende Moment ist gerade das Empirische, das nicht durch Denken und reine Anschauung eingesehen werden kann, das eben nur aus der Beobachtung der Tatsachen gewonnen wird. Vergegenwärtigen wir uns einmal den Inhalt dieser Behauptung. Eine bestimmte Größe der störenden Einflüsse läßt sich gewiß nicht voraussagen; nur daß der Einfluß klein ist, wird behauptet. Aber für den Einzelfall des Experiments ist auch dies nicht mit Gewißheit zu behaupten. Wenn ich angebe, daß für ein bestimmtes fallendes Eisengewicht die Fallzeit nahezu den Wert 2,3 Sekunden haben wird, so ist darin z. B. die Voraussetzung eingeschlossen, daß inzwischen keine größeren unterirdischen Ausbrüche große
61 Mengen von Magnet|eisen unterhalb meines Laboratoriums angesammelt haben. Vorher war der Einfluß ungleicher Verteilung in der Erdmasse gleichfalls vorhanden, nur gehörte er zu den unmeßbar kleinen Einwirkungen; ob aber aus ihm nicht plötzlich ein deutlich meßbarer Faktor wird, darüber kann ich gar nichts aussagen. Mit Gewißheit kann ich jedenfalls nicht verneinen, daß durch eine derartige Veränderung des Kraftfeldes plötzlich starke Abweichungen der Fallzeit entstehen. Es ist auch nicht möglich, diesen Fall in der Voraussetzung auszuschließen; denn wie vorher gezeigt worden ist, besteht eben das Eigentümliche des empirischen Gegenstandes gerade in der Nichterschöpfbarkeit seiner Bestimmungsstücke, und die restlose Ausscheidung aller dieser nicht gemessenen Bedingungen würde den Gegenstand zu einem Gegenstand der reinen Anschauung, das physikalische Urteil zu einem mathematischen reduzieren. Der physikalische Gegenstand ist stets der Gegenstand, wie er mit den Sinnen wahrgenommen wird, ob diese Wahrnehmung nun gerade in diesem Augenblick geschieht oder bloß gedächtnismäßig reproduziert wird; über ihn als Stück der Wirklichkeit wird im physikalischen Sinn etwas ausgesagt, und nicht über die bloße mathematisch-anschauliche Struktur, die für dieses Stück Wirklichkeit Geltung besitzt. Wenn aber wirklich hier eine Aussage über die Größe der ganz beliebigen, nicht gemessenen störenden Einflüße gemacht wird, so läßt sich für keinen Einzelfall – und damit für keine Klasse von Einzelfällen – mit Gewißheit etwas behaupten; jeder beliebige Zahlenwert der Störung ist möglich; und die Behauptung des Beispiels, daß die Fallzeit 2,3 Sekunden betragen wird, muß angesichts der vollkommenen Unwissenheit, vor der wir hier stehen, absurd erscheinen.

Die bloße Annahme der kausalen Verknüpfung alles Naturgeschehens ist deshalb nicht ausreichend, um zu Aussagen über bestimmte wirkliche Dinge –

this constant also has to hold for the instance, and consequently it also has to be true that the measurement of the constant results in a numerical approximation for the particular unique process. Hence, physical experience consists in the establishment of a numerical approximation by experiment, and conversely the physical principle assures by virtue of experience the approximate validity of certain numbers for a class of natural events.

It is a peculiarity of experience that it asserts a rule for effects that are uncontrollable in their diversity and quantity. Everything else about a physical principle consists in mathematical relations; this one additional aspect is the empirical part that can only be obtained by observing facts and not by thought and pure intuition. Let us bring out the point of this claim: One cannot predict the particular size of the interfering influences; we can only say that the influence is small. But not even this can be said with certainty for the individual experiment. If I claim that the time of fall for a particular falling weight of iron will be 2.3 seconds, then, for example that contains the assumption that there has not in the meantime been an accumulation of large quantities of magnetic iron | 61
due to subterranean eruptions below my laboratory. Previously the influence of an uneven distribution of the Earth's mass was also present, but it formed part of the unmeasurably small influences; whether it suddenly turns into a clearly measurable factor is impossible to say. I cannot deny with certainty that such a change in the force field will not result in changes in the time of fall. It is also impossible to exclude this case by assumption since, as has been shown before, the peculiarity of the empirical object consists in the impossibility of grasping all its features. The complete exclusion of all these unmeasurable conditions would make the object one of pure intuition and reduce the physical judgment to a mathematical one. The physical object is always the object as it is perceived by the senses, no matter whether at the moment of perception or when reproduced in the mind. Physical claims are made about this piece of reality, not merely about the simple mathematically intuited structure that is valid for it. However, if a claim is made about the size of an arbitrary unmeasured interfering influence, then nothing can be claimed with certainty about an individual instance – and also not for any class of instances. Any arbitrary numerical value of the interference is possible, and the claim in the example that the drop time is 2.3 seconds must appear absurd given the total lack of knowledge with which we are faced.

The simple assumption of a causal connection between all events in nature is insufficient to make claims about real things – and only those form part of

und nur solche sind Gegenstand der Physik – zu gelangen. Es muß, wenn es physikalische Erkenntnis geben soll, noch ein anderes Prinzip zu dem der Kausalität hinzukommen. Wir müssen nach einem zweiten synthetischen Urteil apriori suchen.

Gäbe es eine restlos genaue Messung der wirklichen Dinge, so müsste sich behaupten lassen, daß der einmal bestimmte Größenwert sich zu jeder Zeit und an jedem Ort wiederfinden lassen müßte. Da die stetige Gleichheit des Größen-
62 wertes nicht behauptet | werden kann, so bleibt uns nichts anderes anzunehmen übrig, als daß, wenn auch nicht dieses, so doch überhaupt ein *Gesetz* für seine Verteilung im Raum und in der Zeit existiert. Dies ist das synthetische Urteil apriori, welches wir fällen müssen. Es ist noch hinzuzufügen, daß das gleiche auch für die Kombination mehrerer Vorgänge gelten muß. Auch für diese läßt sich die Konstanz des berechneten Wertes nicht behaupten, und an ihre Stelle muß die gesetzmäßige Verteilung im Raum und in der Zeit treten.

Wir finden demnach, daß das *Prinzip der gesetzmäßigen Verknüpfung* alles Geschehens, wie sie die Kausalität leistet, nicht zur mathematischen Darstellung der Wirklichkeit ausreicht. Es muß noch ein anderes Prinzip hinzukommen, welches die Ereignisse gleichsam in der Querrichtung, miteinander verbindet; dies ist das *Prinzip der gesetzmäßigen Verteilung.*

Die physikalische Erkenntnis besteht in der Zuordnung von Gleichungen und damit von Zahlen zu Klassen von Gegenständen der empirischen Anschauung. Die Gleichheit dieser Zahlen für eine Anzahl wirklicher Gegenstände der Klasse läßt sich nicht behaupten, sondern nur die näherungsweise Gleichheit; der Sinn dieser Näherung ist der, daß ein Gesetz für die Verteilung der Zahlenwerte besteht. Während mathematische Urteile Größen derart bestimmen, daß diese für alle ihre Einzelgegenstände an jedem Ort und zu jeder Zeit gleich sind, sind die im physikalischen Urteil bestimmten Größen nicht in allen Einzelgegenständen der Klasse gleich, sondern einem Gesetz der Verteilung in Raum und Zeit unterworfen. An Stelle der Allgemeingültigkeit der mathematischen Sätze tritt bei physikalischen Urteilen die Einordnung in das Gesetz der Verteilung.

Über die Gestalt dieses Gesetzes läßt sich noch einiges bestimmen. Es stellt ein Gesetz dar für die Häufigkeit eines Größenwertes bei seiner Wiederholung in der Zeit oder auch bei seiner Vervielfältigung im Raume; daraus folgt, daß diese Häufigkeit nur relativ zu den anderen Häufigkeiten bestimmt sein kann; denn sie muß von der Zahl der Wiederholungen des Vorgangs respektive seiner Vervielfältigungen im Raume überhaupt abhängen und kann keine ein für allemal gültige Konstante sein. Infolge dieser Relativität muß ferner für *jede* Größe der Störungen eine Häufigkeitszahl angebbar sein, denn wäre eine unbestimmt,
63 so wären es damit alle anderen auch. Das Gesetz muß also die Form einer | *Funktion* annehmen, die jedem Wert der Störung eine Häufigkeitszahl zuordnet. (Die Zahl 0 kann auch angenommen werden.) Das Gleiche gilt für die Häufigkeit von Kombinationen, so daß auch Funktionen von mehreren Veränderlichen auftreten müssen.

physics. If we want to have physical knowledge, then a further principle has to be added to causality. We have to search for a second synthetic a priori judgment.

If there were a fully exact measurement of real things, then one should be able to claim that the value obtained once can be found again at any time at any place. Since we cannot claim that the value remains constant | we have no choice 62
but to assume that if it is not constant then there exists some *law* for its distribution in time and space. This is the synthetic a priori judgment which we must make. It should be added that the same judgment must hold for the combination of several processes. As before, we cannot claim that the calculated value remains constant, and we have to replace it with its lawful distribution in space and time.

Hence we conclude that the *principle of lawful connection* of all events, which causality brings about, is insufficient for the mathematical representation of reality. A further principle has to be added, which connects the events – one could say – orthogonally; this is the *principle of lawful distribution.*

Physical knowledge consists in the coordination of equations, and consequently of numbers, to classes of objects of empirical intuition. The equality of these numbers for a number of actual objects of the class cannot be asserted, but only the approximate equality. The reason for this approximation is the existence of a law for the distribution of values. While mathematical judgments determine variables in such a way that they are the same for all their individual objects in all places at all times, the variables in a physical judgment are not equal for all individual objects in their class, but rather subject to a law of distribution in space and time. Instead of the general validity of mathematical claims, we have in the case of physical judgments the subsumption under the law of distribution.

A little more can be said about the form of this law. It is a law for the frequency of each value in repetitions in time or duplicates in space. It follows that the frequency can only be determined relative to other frequencies, since it has to depend on the number of repetitions of the process or its duplicates in space, and cannot be a constant valid for all times. As a result of this relativity, it has to be possible to specify a frequency for *every* size of interference, since if one were undetermined, all others would be too. Consequently, the law must have the form of a | *function*, that assigns to every value of the interference a frequency. 63
(The number 0 may also be assigned.) The same applies to the frequency of combinations, hence functions of several variables must occur.

Es muß jedoch, über diese Funktion noch etwas behauptet werden. Die Gesamtheit der Zahlenwerte, die der störende Einfluß annehmen kann, ist offenbar eine nicht abzählbar unendliche Mannigfaltigkeit; es kann jeder beliebige reelle Zahlenwert angenommen werden. Auch wenn man den Wert zwischen Grenzen einschließen kann, behält die Menge ihre Mächtigkeit. Würde nun jedem Einzelwert eine Häufigkeitszahl zugeordnet, so wäre das Gesetz überhaupt gar nicht realisierbar. Denn nach einer Reihe von Wiederholungen, die stets endlich ist, würden eventuell eine Anzahl von Zahlenwerten, ihren Häufigkeitszahlen entsprechend, oft daran gekommen sein, aber im Verhältnis zu den unendlich vielen anderen Werten wären sie bevorzugt; diese wären nicht ihren Häufigkeitszahlen entsprechend daran gewesen. Es ließe sich niemals eine Verteilung erreichen, die der Forderung des Gesetzes Rechnung trüge, denn über eine endliche Anzahl von Wiederholungen kommen wir nicht hinaus. Das Gesetz aber hat für uns keinen Sinn, so lange wir nicht wenigstens von einer Annäherung an seine Forderung durch endliche Erfahrung reden können; denn nur um der Möglichkeit solcher Erfahrung willen hatten wir seine Existenz gefordert. Deshalb müßen wir die Funktion anders auffassen. Nicht für einmalige Zahlwerte darf sie Häufigkeitszahlen angeben, sondern für Intervalle; dann können wir sagen, daß wenn wir die Gesamtgröße der möglichen Störung als endlich voraussetzen und in $n$ gleich große, sehr kleine, aber doch endliche Intervalle $\Delta x$ teilen, in jedes Intervall eine Anzahl von Werten fallen wird, die durch die Häufigkeitszahl mit Näherung bestimmt ist. Und wir können dann hinzufügen, daß dies schon bei kleineren $\Delta x$ und um so genauer erfüllt sein wird, wenn die Zahl der Wiederholungen wächst, usw., d. h. wir können die Häufigkeitszahlen durch eine integrierbare Funktion $y = \phi(x)$ darstellen, derart daß[E3]

$$N \int_a^b \phi(x)\, dx$$

unter den $N$ Wiederholungen die Zahl derjenigen vorstellt, die in das Intervall $a$ bis $b$ fallen, respektive

64

$$N \int_a^b \dots \int_p^q \phi(x, \dots, x_r)\, dx \dots\, dx_r$$

den entsprechenden Ausdruck für die Funktion von mehreren Veränderlichen. Ist der Spielraum der Störung nicht zwischen endlichen Grenzen eingeschlossen, so muß eine asymptotische Annäherung der Funktion an die $x$-Achse behauptet werden, da sonst wieder nicht durch eine endliche Zahl von Fällen eine Annäherung an die geforderte Verteilung erreicht werden kann; d. h. die Funktion muß mit der $x$-Achse ein endliches Flächenstück abgrenzen.

Diese Funktion ist aber genau die früher entwickelte Wahrscheinlichkeitsfunktion.

E3. The original German contains an error, there the integral is: $N \int_a^b \phi(x)\, dy$

However, one further claim has to be made about the function. The totality of numbers that the interfering influence may assume is clearly an uncountably infinite manifold; it can take any arbitrary real number. Even if one restricts the value between limits, the set retains its cardinality. If one were to assign a frequency to every individual value, then the law would not even be realizable. After a finite sequence of repetitions, a number of values might have occurred corresponding to their frequency, but in comparison to the infinitely many other values that did not occur in proportion to their frequency they would be privileged. We could never obtain a distribution that would satisfy the requirement of the law, since we will never have more than finitely many repetitions. But the law is no use to us unless we can at least speak of an approximation of its requirements by finite experience; after all we have only required its existence in order to have such experience. Hence we have to conceive of the function differently. It should not assign frequencies to individual values, but to intervals; then we can say that if we assume the total size of the possible interference to be finite and divide it into $n$ equal-sized, very small but finite intervals $\Delta x$, then a number of values, approximately determined by the frequency, will fall in each. And we can add that this is more precisely satisfied for small $\Delta x$ as the number of repetitions increases, i. e., we can represent the frequencies by an integrable function $y = \phi(x)$ such that

$$N \int_a^b \phi(x)\, dx$$

represents those among the $N$ repetitions that fall into the interval $a$ to $b$, or
correspondingly 64

$$N \int_a^b \ldots \int_p^q \phi(x, \ldots, x_r)\, dx \ldots dx_r$$

represents the expression for the function of several variables. If the range of the interference is not bounded by finite limits, then one has to assume that the function asymptotically approaches the $x$-axis, since otherwise we again cannot achieve an approximation of the required distribution in a finite number of cases; i. e., the function must bound a finite area with the $x$-axis.

This function is just the probability function developed earlier.

## 65 4. Die Stellung der Wahrscheinlichkeitsurteile zur Wirklichkeit

**I.**

Es ist damit die Existenz einer Wahrscheinlichkeitsfunktion deduziert worden in dem Sinne, wie Kant das Wort Deduktion für die Transzendentalphilosophie gebraucht. Die Notwendigkeit einer solchen Gesetzmäßigkeit läßt sich letzten Endes nur einsehen, und sie ist deshalb ein synthetisches Urteil a priori genannt worden; sie läßt sich nicht logisch aus anderen Grundsätzen der Erkenntnis ableiten. In dieser Deduktion aber ist gezeigt worden, wie jenes Gesetz im Zusammenhang steht mit der gesamten Naturerkenntnis überhaupt; und indem dargetan worden ist, daß jenes Prinzip eine notwendige Bedingung aller physikalischen Erkenntnis bedeutet, ist seine Gültigkeit von der Erfahrung *bewiesen* worden.

Die Behauptungen, die die Wahrscheinlichkeitsrechnung aufstellt, gelten deshalb mit Gewißheit von den Gegenständen der Wirklichkeit. Mit Gewißheit, nicht wieder bloß mit Wahrscheinlichkeit; denn irgendeine Gesetzmäßigkeit muß mit Gewißheit gelten, wenn andere mit Wahrscheinlichkeit gelten sollen. Wahrscheinlich ist es, daß der einzelne Fall dem Maximalwert der Kurve $\phi(x)$ entspricht; gewiß ist es, daß mit wachsender Zahl der Fälle eine Annäherung an eine solche Kurve zustande kommt. Und wenn sich in einer Reihe von Versuchen diese Annäherung nicht zeigt, so werden wir annehmen müssen, daß die
66 Bedingungen einer Wahrscheinlichkeitsverteilung nicht gegeben waren, nicht | aber, daß das Gesetz falsch ist. Es handelt sich hier um ein metaphysisches Prinzip der Naturerkenntnis. Aufgabe der Erfahrung ist es nur, seinen jeweiligen Inhalt im speziellen Falle zu bestimmen; d. h. die besondere Gestalt der Funktion $\phi(x)$ jeweils zu ermitteln, z. B. ob sie das Gaußsche Gesetz darstellt oder anderen Inhalt hat. Analog zum Prinzip der Kausalität stellt das Prinzip der Verteilung nur die allgemeine Form vor, in die spezielle Erfahrung einen speziellen Inhalt hineinfügt. Deshalb muß auch jeder Versuch, das Prinzip durch die Erfahrung zu bestätigen, dem Philosophen ebenso lächerlich erscheinen, wie ein etwaiger Versuch der experimentellen Prüfung des Kausalprinzips. Es ist bezeichnend, daß Forscher, die ganz ernsthaft den Versuch gemacht haben, das Wahrscheinlichkeitsgesetz experimentell zu beweisen, doch gar nicht fähig waren, sich seinem aprioren Zwange zu entziehen. So hat R. Wolf in dieser Absicht mit Würfeln 120000 Würfe ausgeführt; als aber bei einem Würfel die Verteilung gar nicht stimmen wollte, hat er lieber auf eine exzentrische Lage des Schwerpunktes geschlossen als auf die Nichtgültigkeit des Wahrscheinlichkeitsgesetzes.[23] Das einzige Resultat, was er durch seine 120000 Würfe demnach erreicht hat, ist die Entdeckung, daß sein einer Würfel nicht gleichmäßig gebaut war; weiter hat dieser Versuch für die Wissenschaft nichts bewiesen. Es erscheint aber zweifelhaft, ob dieses Resultat eine dem Aufwand seiner Gewinnung entsprechende Bereicherung der Wissenschaft bedeutet.

23. Vgl. Czuber, *Wahrscheinlichkeitsrechnung*, S. 149.

## 4. The Relation of Judgments of Probability to Reality 65

**I.**

We have deduced the existence of a probability function in the same sense as Kant uses the word deduction for transcendental philosophy. Ultimately, the necessity of such a law must be recognized, and hence we refer to it as a synthetic a priori judgment; it cannot be derived logically from other principles of knowledge. But in this deduction it has been shown how such a law is connected to knowledge of nature in general; and by showing that such a principle is a necessary condition for all physical knowledge, its validity for experience has been *proven*.

Consequently, the assertions probability theory makes are true with certainty for the objects of reality. With certainty, not again just with probability; because some law has to hold with certainty if others are to hold with probability. It is probable that the individual case corresponds to the maximum of the curve $\phi(x)$; it is certain that with an increasing number of instances we get an approximation to such a curve. And if we do not find such an approximation in a sequence of trials, then we have to assume that the conditions for the probability distribution were not satisfied, but we do not have to conclude | that the law 66
is wrong. We are dealing with a metaphysical principle of knowledge of nature. The only task of experience is to determine its respective content in a specific case; i. e., to determine the particular shape of the function $\phi(x)$; whether for example it represents the Gaussian law or has a different content. Analogously to the principle of causality, the principle of distribution only describes the general form that specific experience fills with specific content. Hence any attempt to verify the principle by experience must seem just as futile to any philosopher as any attempt to experimentally test the principle of causality. It is notable that researchers who seriously attempted to prove the principle of probability experimentally were unable to escape its a priori force. For example, R. Wolf performed 120,000 throws of dice with this intention; but when one of the dice was anomalously distributed, he concluded that the center of gravity was in an eccentric position rather than concluding that the law of probability is invalid.[23] Hence the only result he obtained with his 120,000 throws was the discovery that one of the dice was not constructed evenly; it did not prove anything further for science. It seems doubtful whether this result advanced science in proportion to the effort required to obtain it.

23. See Czuber, *Wahrscheinlichkeitsrechnung*, p. 149.

## II.

Wir können den Inhalt des Wahrscheinlichkeitsprinzips so aussprechen: Es existiert für eine Reihe von Wiederholungen oder Vervielfältigungen derselben Größe eine Wahrscheinlichkeitsfunktion. Damit ist nicht verlangt, daß diese Größe immer durch denselben Vorgang dargestellt sein muß. So werden die Zahlenwerte der Gaskonstanten $R$ die Verteilung zeigen, auch wenn die Messung an verschiedenen Gasen in ganz verschiedenen Zuständen vorgenommen würde. Es stellt einen besonderen Fall vor, wenn stets nur „derselbe" Vorgang wiederholt wird; hiermit ist meist eine Klasse von solchen Fällen ausgewählt, die sich
67 nur unterhalb der Grenzen | der Meßbarkeit unterscheiden. Aber die durch unsere jeweiligen Meßinstrumente gesetzte Grenze ist recht willkürlich, und es wäre deshalb ungereimt, die behauptete Gesetzmäßigkeit der Verteilung auf Einflüsse zu beschränken, die unterhalb der zufälligen Messungsgrenze liegen; so kommt es, daß wir auch noch von Wiederholung „desselben" Vorgangs sprechen müssen, wenn die Schwankungen bereits deutlich meßbar sind, etwa in der Stärke der Stöße bei der würfelnden Hand. Auch für die in solchen Vorgängen wiederholten Größen gilt das Gesetz der Verteilung.

In dem besonderen Falle, wo durch äußere Vorrichtungen diese Größenwerte in Scharen gleicher Intervalle klassifiziert werden, wird die Anzahl der Wiederholungen für jede Schar nahezu gleich werden; wie dies aus der Existenz der Wahrscheinlichkeitsfunktion mit mathematischer Strenge folgt, ist im ersten Abschnitt des zweiten Kapitels dargetan worden. Die Gültigkeit der Wahrscheinlichkeitsverteilung für Fälle, in denen scheinbar von einer Wahrscheinlichkeitsfunktion nicht die Rede ist, sondern wo es sich um Zählung von Häufigkeiten bestimmter Klassen handelt, ist deshalb durch die gegebene Deduktion gleichfalls bewiesen worden.

Es muß hier jedoch noch von einem Paradoxon gesprochen werden, das man angeführt hat, um dem Wahrscheinlichkeitsgesetz den Anspruch der Gewißheit strittig zu machen. Für einen Würfel haben alle Seiten die gleiche Wahrscheinlichkeit, getroffen zu werden; soll nun mit Gewißheit eine Aussage über die Häufigkeit ihres Vorkommens gemacht werden, so wird man sagen müssen, daß etwa unter 600 Fällen jede Seite nahezu 100mal darankommt; man wird noch Genauigkeitsgrenzen hierfür angeben können. Diese Aussage ist dann aber gewiß, und es ist daher *unmöglich*, mit 600 Würfen eine Verteilung zu erreichen, in der die Seite 1 nur 10mal vorkommt. Eine andere Betrachtung jedoch führt zum Gegenteil; betrachte ich die 600 Würfe als *einen* Fall, als Element einer neuen Reihe von 100000mal 600 Würfen, so ist unter den 600 Würfen jede Kombination gleich wahrscheinlich, also der Fall, daß unter den 600 Würfen die 1 nur 10mal darankommt, keineswegs unmöglich. Man setzt dann jede verschiedene Anordnung der 600 Würfe als gleich wahrscheinlich und zählt die Zahl der Kombinationen, die die 1 10mal enthalten; entsprechend dieser Häufigkeitszahl ist dann die Wahrscheinlichkeit des Falles, daß die 1 10mal darankommt. Ja, es
68 muß sogar behauptet werden, daß, | wenn ich die neue, höhere Mannigfaltigkeit nur groß genug wähle, sich sogar mit Gewißheit der Fall der 10 Einsen einstellen wird, denn es soll ja bei genügender Wiederholung mit Gewißheit eine Ver-

## II.

We can formulate the content of the principle of probability in the following way: There exists a probability function for a sequence of repetitions or duplicates of the same quantity. This does not require that this quantity has to be represented by the same process each time. For example the values of the gas constant $R$ will show a distribution even when the measurements are made on different gases in completely different states. It is a special case if the "same" process is repeated each time; in this case we are generally considering a class of instances that only
differ below the limits | of what can be measured. But the limit given by our re- 67
spective measurement instruments is quite arbitrary, and consequently it would be misleading to limit the asserted law of distributions to influences that are below the limit of what can be measured; that is why we have to refer to repetitions of the "same" process even if the variations are already clearly measurable, as is the case with the strength of tosses of the hand that throws the dice. The law of distribution also applies to the repeated quantities in these processes.

In the special case where the values of variables are classified in sets of equal intervals by external devices, the number of repetitions of each set will be almost equal; how this follows with mathematical exactness from the existence of a probability function has been shown in the first part of the second chapter. The previous deduction also proves the validity of probability distributions for cases where there does not appear to be any reference to probability functions, but rather where we are counting the frequencies of particular classes.

We still have to discuss a paradox that has been raised to dispute the claim that the law of probability is certain. In the case of a die all sides have the same probability of coming up; if one wants to make a claim with certainty about the frequency of their occurrence, then one will have to say that among 600 throws, each face will appear approximately 100 times; one will be able to specify bounds of precision. This claim is certain, and it is therefore *impossible* to obtain a distribution with 600 throws in which the face with "1" will only appear 10 times. However, a different approach leads to the opposite conclusion; consider the 600 throws as *one* instance, as an element of a new sequence of 100,000 times 600 throws; then among the 600 throws every combination is equally possible and consequently the case where among the 600 throws the "1" only appears 10 times is by no means impossible. Assume each sequence of 600 throws to be equally likely and count the number of combinations that contain "1" 10 times; the probability of "1" appearing 10 times corresponds to this frequency. Indeed,
we even have to assert that | when the new, greater manifold is chosen suffi- 68
ciently large, the case of 10 "1"s must occur with certainty, since with a sufficient number of repetitions a distribution should arise with certainty, in which

teilung sich herstellen, in der jeder Fall seiner Wahrscheinlichkeit entsprechend darankommt. Hier muß also behauptet werden, daß der Fall, mit einem Würfel unter 600 Würfen nur 10mal die 1 zu treffen, mit Gewißheit eintreten wird, während er in der ersten Betrachtung als unmöglich erschienen war.

Der Widerspruch löst sich auf folgende Weise. In der ersten Reihe war die 1 nahezu 100mal daran gekommen. In der neuen, höheren Mannigfaltigkeit kommt zwar der Fall vor, daß die 1 unter 600 Würfen nur 10mal getroffen wird, aber viel häufiger auch der Fall, daß die 1 100mal und mehr getroffen wird. Zählt man nun in der höheren Mannigfaltigkeit durch, so wird man finden, daß unter den $100000 \times 600$ Würfen die 1 wieder nahezu in $\frac{1}{6}$ der Fälle getroffen wurde, mit viel größerer Genauigkeit sogar als bei den 100 Fällen. Die Bestimmung dieser Verteilung ist ein mathematisches Problem; es lautet so: wenn ich sämtliche Kombinationen, die unter 600 Würfen möglich sind, nebeneinander schreibe, wie oft schreibe ich dann die Zahl 1 nieder? Bernoulli hat in seinem bekannten Theorem die Antwort darauf gegeben; er hat exakt bewiesen, daß in der Gesamtheit der so niedergeschriebenen Zahlen jede der 6 Ziffern genau gleich oft daran kommt. Die Zahl wird übrigens sehr groß; es werden dann im ganzen $600 \times 6^{600}$ Ziffern niedergeschrieben sein.

Die Behauptung, daß im ganzen jede Seite gleich oft daran kommt, steht also nicht im Widerspruch zu der anderen Behauptung, daß auch außergewöhnliche Reihenfolgen gelegentlich vorkommen. Die ganze Paradoxie war nur durch einen Fehler hereingekommen, der in die Formulierung eingegangen war. Man hatte nämlich behauptet, daß in der bestimmten Zahl von 600 Würfen die Gleichverteilung mit einer bestimmten, vorgegebenen Genauigkeit erreicht würde. Dies aber gerade kann das Wahrscheinlichkeitsprinzip nicht aussagen. Es kann nur besagen, daß überhaupt einmal eine Gleichverteilung erreicht wird und daß eine Annäherung dahin stattfindet; aber mit welcher bestimmten Zahl von Würfen dies geschieht, darüber vermag es nichts anzugeben.
69 Denn diese Zahl hängt eben gerade nicht von den gemessenen | Bedingungen des Vorgangs, sondern von den *ungemessenen* ab; man wird sagen können, daß bei größerer Präzision einer Vorrichtung die Gleichverteilung eher erreicht wird, so daß umgekehrt die Zahl der Reihe ein Maß der störenden Einflüsse wird. Es läßt sich nicht einmal behaupten, daß bei Wiederholung des Versuchs mit demselben Würfel die Gleichverteilung mit annähernd der gleichen Anzahl erreicht wird, denn es lassen sich eben nur die meßbaren Bedingungen wieder ebenso wie früher herstellen, gerade die ungemessenen können ganz andere geworden sein. Darum ist es falsch, jemals behaupten zu wollen, unter 600 Würfen wird mit Gewißheit diese oder jene Genauigkeit der Gleichverteilung erreicht. Wir können nur sagen, daß überhaupt eine solche Annäherung zur Gleichverteilung stattfinden wird; es ist ein wichtiges Verdienst des Bernoullischen Theorems, nachgewiesen zu haben, daß dies keinen Widerspruch zu dem Vorkommen der ungewöhnlichsten Reihenfolgen bedeutet.

Über die Reihenfolge der einzelnen Fälle ist deshalb nichts ausgesagt. Es ist nicht ausgeschlossen, daß eine Reihe von Würfen mit zwanzigfacher Wiederholung der 1 beginnt, dann wird erst in späteren Wiederholungen dieses Überwie-

each case occurs according to its probability. Consequently, we have to claim that throwing a "1" only 10 times with a die in 600 throws must occur with certainty, while previous considerations seemed to suggest it was impossible.

The contradiction is solved in the following way. In the first sequence, the "1" occured approximately 100 times. In the new, greater manifold the case that the "1" is thrown 10 times in 600 throws occurs, but the case that the "1" is thrown 100 times or more occurs much more often. If one now counts up in the greater manifold, then one will find that among the $100,000 \times 600$ throws the "1" will again occur in close to $\frac{1}{6}$ of the cases – in fact, with much greater precision than with the 100 cases. Determining this distribution poses the following mathematical problem: if I write next to each other all combinations that are possible in 600 throws, how often will I write down the number "1"? Bernoulli gives the answer in his famous theorem: he proved precisely that in the entirety of the numbers written like this, each of the 6 digits will appear exactly equally often. In fact the number is going to be very large; in total there will be $600 \times 6^{600}$ digits.

Hence the assertion that in total each side will appear equally often does not contradict the other claim that unusual series occur occasionally. The whole paradox only arose because of a mistake that was made in the formulation. It was claimed that the uniform distribution would be achieved with a particular given precision in the fixed number of 600 throws. But this is exactly the claim that the principle of probability cannot make. It can only claim that a uniform distribution will be reached eventually and that an approximation towards it occurs; but how many throws are required it cannot specify. Because this number does not depend on the measured conditions of the process, | but depends on the *unmeas-* 69
*ured* ones, one can say that devices of greater precision reach the uniform distribution sooner; conversely, the number of throws in the series becomes a measure for the interfering influences. One cannot even claim that a repetition of the experiment with the same die will reach the uniform distribution with roughly the same number of throws, because only the measurable conditions can be reconstructed; the unmeasured conditions may have changed completely. Hence it is false to claim that a particular approximation of the uniform distribution is reached with certainty in 600 throws. We can only say that such an approximation of the uniform distribution will occur at some point; it is an important achievement of Bernoulli's theorem to have shown that this does not contradict the occurrence of the most unusual sequences.

Hence, nothing is said about the order of succession in an individual sequence of throws. It is not ruled out that a sequence of throws starts with 20 repetitions of a "1", which will only be balanced out by later repetitions. But again

gen wieder ausgeglichen werden. Wieder aber wäre es verkehrt, daraus schließen zu wollen, daß, nachdem die 1 erst zwanzigmal aufgetreten ist, ihr ferneres Auftreten nun unwahrscheinlicher wäre als das der anderen Zahlen. Dies würde allerdings ein Gesetz der Reihenfolge besagen. Aber es ist falsch, zu glauben, daß die behauptete Annäherung nur durch weniger häufiges Auftreten der 1 erreicht werden kann. Auch wenn von nun ab jede Zahl gleich oft daran käme, wird mit der Anzahl der Versuche der anfängliche Überschuß der 1 von immer kleinerem Einfluß werden und eine Annäherung an die Gleichverteilung stattfinden. Dies gilt aber nicht nur für 20, sondern für jede beliebige endliche Zahl. Ausgeschlossen wird durch die Wahrscheinlichkeitsbehauptung nur der Fall, daß ein *dauerndes* überwiegen der 1 stattfindet, daß also die Anzahl der Wiederholungen nach einem anderen Verhältnis als dem der Gleichverteilung hinstrebt.

Es wäre jedoch irrtümlich, zu glauben, daß, weil ein bestimmtes $N$, mit dem die verlangte Approximation eintritt, nicht angegeben werden kann, das Prinzip der Verteilung damit inhaltslos geworden wäre. Die Behauptung, es gibt ein end-
70 liches $N$ | derart, daß es die Abweichung von der geforderten Verteilung kleiner als die beliebig vorgegebene Zahl $\epsilon$ macht, hat einen klaren und bestimmten Sinn auch ohne daß dies $N$ angegeben werden kann, und schreibt der Wirklichkeit ein besonderes Verhalten vor. Zwar kann ich ihr Gegenteil in keiner Erfahrung konstatieren, denn wenn die Abweichung bei irgendeinem $N$ noch größer ist als $\epsilon$, so bleibt stets der Ausweg offen, daß dieses $N$ eben noch zu klein war. Das ist aber mit keinem aprioren Prinzip anders. Auch eine Abweichung vom Kausalgesetz könnte niemals konstatiert werden, sondern jedesmal müßte die Erklärung der betreffenden Beobachtung so gefaßt sein, daß sie das Verhalten doch wieder als ein kausales, nur anderer spezieller Beschaffenheit, darstellt. Das ist gerade das Kriterium apriorischer Gesetze, daß sie nicht durch irgendeine spezielle Erfahrung bestätigt oder widerlegt werden können, sondern die vorher gesetzten Formen der Einordnung bilden, die erst die spezielle Erfahrung möglich machen. Erfahrung in wissenschaftlichem Sinne ist eine solche Darstellung der Wirklichkeit, die die gegebenen Wahrnehmungsinhalte im Sinne fester apriorischer Ordnungsformen zusammenfügt. Würde das Gesetz der Verteilung für die gewünschte Approximation ein bestimmtes $N$ angeben können, so würde es eben dadurch seinen apriorischen Charakter verlieren und zu einem speziellen Naturgesetz, welches das Experiment bestätigen oder widerlegen kann, herabsinken.

Eine ernstere Schwierigkeit ähnlicher Art entsteht jedoch durch folgende Überlegung. Wir hatten die Approximation dadurch definiert, daß wir sagten, es gibt ein endliches $N$ derart, daß die Abweichung in den beliebig vielen $r$ Wahrscheinlichkeitsfunktionen gleichzeitig kleiner als $\epsilon$ wird. Was mit weiter wachsendem $N$ geschieht, ob die Abweichung wieder größer wird oder nicht, darüber können wir nichts aussagen; wir wissen nur, daß es ein noch größeres $N$ gibt, derart, daß die Abweichung noch beliebig kleiner als $\epsilon$ wird. Man mag nun einwenden, dadurch aber werde das Prinzip leer; denn wenn die Abweichung jederzeit wieder beliebig groß werden kann, so kann man ebenso gut sagen: es gibt ein $N$ derart, daß irgendeine beliebige Dispersion erreicht wird. Dann aber

it would be wrong to conclude that now that the "1" has appeared 20 times, it is less likely than other numbers to appear later. This is what would be claimed by a law of succession. But it is wrong to believe that the postulated approximation can only be obtained by fewer occurrences of the "1". Even if from now on every number should occur equally often, with an increasing number of trials the initial excess of "1"s will have less and less influence and an approximation of the uniform distribution will occur. This does not only hold for 20, but equally for any arbitrary finite number. The only case excluded by the probability claim is a *persistent* excess of "1"s, i. e., that the number of repetitions tends to a ratio other than the equal distribution.

It would be mistaken to believe that the principle of distribution is void of content merely because a particular $N$ at which the desired approximation occurs cannot be specified. The claim that there is a finite $N$ | such that the de- 70
viation of the described distribution is smaller than an arbitrary given number $\epsilon$ has a clear and definite meaning even if this $N$ cannot be specified, and it imposes a special constraint on reality. It is true that no experience can establish its denial, since if a deviation is larger than $\epsilon$ for some $N$ then there is always the solution that this $N$ is still too small. But this is no different with any other a priori principle. A deviation from the law of causality can also never be established; in each case the explanation of the respective observation could be so presented that the behavior is causal after all, just of a different specific kind. The criterion of a priori laws is that they cannot be confirmed or refuted by any specific experience, but rather constitute the prior forms of classification that make the specific experience possible. In its scientific sense experience is a representation of reality that combines given contents of perception through fixed a priori forms of classification. If the law of distribution were able to specify an $N$ for the desired approximation, then it would lose its a priori character and would be reduced to a particular law of nature which can be confirmed or refuted by experiments.

A more serious complication of a similar kind arises from the following consideration. We had defined the approximation by saying that there is a finite $N$ such that the deviation in the arbitrary $r$ many probability functions[E8] is simultaneously smaller than $\epsilon$. Whether the deviation increases again or not as $N$ grows, we cannot say; we just know that there is an even larger $N$ such that the deviation is arbitrarily smaller than $\epsilon$. One may object that this makes the principle empty since if the deviation may increase arbitrarily again, then one may as well say: there is an $N$ such that any arbitrary distribution is attained. But then

E8. See page 36 [89].

würde man auf Grund des Prinzips alle möglichen Verteilungen behaupten können, und das Prinzip würde dann nichts Bestimmtes mehr besagen. Der Fehler
71 liegt | jedoch in der Annahme, daß die Abweichung wieder beliebig *groß* werden *muß*. Mit *Gewißheit* kann man nur aussagen, daß sie beliebig *klein* werden muß, aber nicht, daß sie irgendeine bestimmte geforderte Abweichung je erreicht, geschweige denn größer als sie wird. Dies läßt sich so formulieren: Mit Gewißheit läßt sich aussagen, daß man aus der Folge der Abweichungen eine nach 0 konvergente Teilfolge herausgreifen kann. Über divergente Teilfolgen läßt sich nichts aussagen.

Es muß, um dies zu klären, zwischen zwei Formen der Behauptung unterschieden werden. Wenn wir irgendeine bestimmte Reihenfolge mit der unnatürlichsten Dispersion innerhalb von $r$ Wiederholungen angeben, so müssen wir allerdings, indem wir den Satz von $r$ Wiederholungen zum Element innerhalb einer großen Serie von $S$ Wiederholungen machen, sagen; es gibt ein $S$ derart, daß mindestens einer von den Sätzen die angegebene Dispersion besitzt. Indem wir die Aussage auf einen *Teil* der Gesamtheit beziehen, können wir ihr den Charakter der Gewißheit verleihen, und dies für jede beliebige Form der Verteilung. Die andere Form ist die, daß wir die Aussage auf die *Gesamtheit* beziehen. Hier können wir nur eine einzige Aussage mit *Gewißheit* machen: es gibt ein $N$ derart, daß in der Folge von den beliebig vielen $r$ Wahrscheinlichkeitsfunktionen die Abweichung überall unter das beliebig kleine $\epsilon$ sinkt. Damit erscheint die eine Verteilung, die der normalen Dispersion, vor den anderen ausgezeichnet: daß sie angenommen wird, können wir von der Gesamtheit der Reihe mit Gewißheit behaupten, daß irgendeine andere angenommen wird, läßt sich dagegen nur von Teilen der Reihe behaupten. Diese Sonderstellung der normalen Verteilung ist aber nicht irgendwie unnatürlich. Denn sie besagt im Grunde dasselbe wie die andere Behauptung, daß jede beliebige Verteilung, als Element einer höheren Serie aufgefaßt, mit Gewißheit daran kommen wird; sie spricht für die Gesamtheit der Reihe dasselbe aus, was die andere Behauptung nur für Teile aussagen kann.

Das Prinzip der Verteilung stellt deshalb ein objektives Gesetz des Naturgeschehens dar, das mit Gewißheit gilt. Es kann nicht für den einzelnen Fall und nicht für eine bestimmte vorgegebene Zahl von Wiederholungen – diese beiden Fälle unterscheiden sich nur graduell – eine bestimmte Verteilung mit
72 Ge|wißheit voraussagen; sondern mit Gewißheit gilt nur, daß es überhaupt eine endliche Zahl $N$ gibt, derart, daß in den vorgegebenen $r$ Funktionen der Folge die Abweichungen kleiner als ein vorgegebenes $\epsilon$ werden. Deshalb, weil es ein solches endliches $N$ gibt, haben wir das Recht, den Eintritt einer gewünschten Annäherung zu erwarten; dies ist eine *vernünftige Erwartung*, weil sie in einem objektiven Gesetz des Geschehens ihren Grund hat. Ihr Inhalt wird durch das Wahrscheinlichkeitsurteil ausgesprochen.

the principle would justify any possible distribution and the principle would no
longer say anything specific. The mistake however lies | in the assumption that 71
the deviation *has* to become arbitrarily *large* again. With *certainty* we can only assert that it must get arbitrarily *small*, but not that it reaches or even exceeds some stipulated deviation. This can be formulated as follows: It can be claimed with certainty that in the sequence of deviations there is a subsequence that converges to 0. No claim can be made about divergent subsequences.

To clarify this matter, one has to distinguish between two forms of assertions. If we specify any particular sequence of $r$ repetitions with the most unnatural dispersion, then, in claiming the $r$ repetitions are a subsequence of a large series of $S$ repetitions, we have to say: there is an $S$ such that at least one of the subsequences has the specified dispersion. By relating the claim to a *part* of the whole, we can assign it certainty, and do so for any form of distribution. The other form is that we relate the claim to the *whole*. Then we can only make one claim with *certainty*: there is an $N$ such that in the sequence of the arbitrary $r$-many probability functions the deviation drops everywhere to below the arbitrary small $\epsilon$. Then one distribution, namely that with the true dispersion,[E9] appears distinguished: we can claim with certainty that the entire sequence will attain it, but that any other distribution is attained we can only claim for parts of the sequence. This privileged position of the true distribution is not somehow unnatural. Ultimately, it amounts to the previous claim that each distribution, seen as an element of a higher series, will occur with certainty; it asserts for the whole sequence what the other claim only asserts for its parts.

Hence the principle of distribution is an objective law of natural events that holds with certainty. It cannot predict for the individual case nor for a particular given number of repetitions – these two cases only differ as a matter of degree –
a particular distribution with certainty; | certainty applies only to the fact that 72
for any given $\epsilon$ there is a finite number $N$ such that in the given $r$ functions the sequence of deviations become smaller than $\epsilon$. Hence, because there is such a finite $N$, we are justified to expect the occurrence of a desired approximation; this is a *rational expectation* because it is based on an objective law of events. Its content is specified by the probability judgment.

E9. In order to avoid confusion with the Normal distribution here, we translated Reichenbach's "*normale Dispersion*" with "true dispersion".

## III.

Die Stellung der Wahrscheinlichkeitsurteile zur Wirklichkeit, die Fick in der Seite 13 zitierten Äußerung als Problem aufgezeigt hatte, ist mit den hier gegebenen Untersuchungen geklärt worden. Wir können unser Resultat im Anschluß an Fick zusammenfassen:

1. Fick hat gezeigt, daß die Sätze der Wahrscheinlichkeitsrechnung inbezug aufeinander synthetische Sätze apriori sind und ein System vorstellen, das den Sätzen der Geometrie analog ist.

2. In dieser Untersuchung wurde gezeigt, daß diese Sätze mit Notwendigkeit von der Wirklichkeit gelten müssen; d. h. daß sie Sätze über die Wiederholung von Ereignissen darstellen, denen sich die wirklichen Dinge notwendig unterordnen. Diese Unterordnung geschieht auf Grund des Prinzips der Wahrscheinlichkeitsfunktion, welches ein objektives Gesetz des Naturgeschehens darstellt.

Erst dadurch, daß dieses zweite Resultat bewiesen wurde, ist die Parallele mit den geometrischen Sätzen vollkommen geworden. Denn deren eigentümlicher Charakter besteht nicht nur darin, daß sie unter sich ein geschlossenes System widerspruchsfreier Urteile darstellen, sondern er liegt erst darin enthalten, daß diese Urteile auch auf die Welt des wirklichen Geschehens anwendbar sind. Nur darum enthalten diese Sätze *Erkenntnisse*, und stellen mehr dar als ein willkürliches Spiel des Verstandes, das zwar den Gesetzen der Logik Rechnung tragen, aber in seiner Gesamtheit nur den komplizierten Ausbau beliebig gesetzter Annahmen bedeuten würde. In eben diesem Sinne würden die Wahrschein-
73 lichkeitsurteile „ein Spiel des vergleichenden Witzes" | sein, wenn es nicht wirkliche Dinge gäbe, die ihnen notwendig untergeordnet sind.

Wie diese Dinge beschaffen sein müssen, damit auf sie die Wahrscheinlichkeitsgesetze Anwendung finden können, ist gleichfalls dargestellt worden. Es ist ausgeführt worden, daß überall, wo eine konstante Größe in einem Vorgang häufig wiederholt verwirklicht wird, die Messungswerte dieser Größe gemäß dem Gesetz einer integrierbaren Funktion, die mit der Abszissenachse ein endliches Flächenstück einschließt, sich verteilen müssen. Durch eine besondere Anordnung, die die Werte entsprechend gleichen Intervallen des Arguments dieser Funktion in Scharen klassifiziert, nimmt dieses Gesetz eine besondere Form an; es ist die Form, die den Sätzen der mathematischen Wahrscheinlichkeitsrechnung gewöhnlich zugrunde gelegt wird und in dem Schema des Würfels zum Ausdruck kommt. Da schließlich in jeder Reihe von Naturvorgängen eine Größe nahezu konstant erhalten bleibt, läßt sich für diese und also für alle Naturvorgänge das Wahrscheinlichkeitsprinzip anwenden. Erst indem dieses Prinzip zum Kausalprinzip hinzukommt, gewissermaßen in der Querrichtung die Ereignisse verbindend, wie es das Verhältnis der Ursache zur Wirkung in der Längsrichtung getan hatte, wird Naturerkenntnis möglich.

Es muß jedoch gegenüber den geometrischen Sätzen ein Unterschied in den Wahrscheinlichkeitssätzen betont werden. Die ersteren stellen Beziehun-

## III.

The relation of judgments of probability to reality that Fick raised as a problem in the quotation on page 13 [57] has been solved by the analysis given here. Following Fick, we can summarize our result:

1. Fick showed that the statements of probability calculus stand in a synthetic a priori relation to one another and present a system that is analogous to the theorems of geometry.

2. We showed in this analysis that these theorems have to be necessarily true of reality, i. e., that they are theorems about the repetition of events to which real things are necessarily subordinate. This subordination is due to the principle of the probability function, which is an objective law of natural events.

Only the proof of this second result completed the parallel to geometrical theorems, since their particular character does not only consist in the fact that they form a closed system of consistent judgments, but rather consists in the fact that these judgments are applicable to the world of real events. This is the only reason why these theorems contain *knowledge* and are more than an arbitrary game of the mind that is subject to the laws of logic but in its entirety only consists in the complicated extension of arbitrarily chosen assumptions. In this sense probability judgments would be "ein Spiel des vergleichenden Witzes"[E10] | if there were no real things that are necessarily subordinate to them. 73

We also described how these things have to be constituted so that the laws of probability are applicable to them. We explained that in every case where a constant variable is realized repeatedly in a process, the measurements of this variable have to be distributed according to the law of an integrable function that bounds a finite area with the $x$-axis. This law takes a special form due to a particular arrangement that classifies the values into sets according to equal intervals of the argument of the function; this is the form on which the theorems of mathematical calculus of probability are usually based, and which was illustrated in the schema of the dice example. Since there is in every sequence of natural processes one variable that remains almost constant, the principle of probability can be applied to this and hence to all other natural processes. Natural knowledge is only possible when this principle is added to the principle of causality, thereby, so to speak, connecting events orthogonally to the direction in which the relation of cause and effect connects them.

However, one difference between the geometrical theorems and theorems of probability has to be emphasized. The former describe relations between ideal

E10. We do not know what this phrase means or what it references. It appears to be a colloquial saying of the time. Literally, it would be translated as "a game of the comparative joke".

gen zwischen idealen Gegenständen vor, die zwar in der Natur selten realisiert, doch auf die Natur mit Näherung anwendbar sind. Der ideale Gegenstand aber ist *vorstellbar*, und es ist keineswegs ausgeschlossen, daß ein wirklicher Körper einmal genau die Gestalt eines einfachen dieser Idealgegenstände besitzt. Ja, es muß sogar behauptet werden, daß jeder Körper, wenn wir seine Gestalt auch niemals vollständig kennen, doch in jedem Zeitmoment eine ganz bestimmte räumliche Form besitzt, und daß wir, wenn wir nur diese Form kennten, bestimmte Gesetze über sie mit Gewißheit aussagen könnten. Die idealen Gegenstände der Geometrie sind in jedem Augenblick verwirklicht, wenn wir sie auch im einzelnen nicht kennen. Bei den Wahrscheinlichkeitssätzen ist es grundsätzlich anders. Es gibt kein Geschehen, das den Idealfall der Wahrscheinlichkeitsrech-
74 nung vorstellt. Denn so oft ich einen Vorgang | auch wiederhole, die Häufigkeiten der Größenwerte werden stets eine diskrete Folge vorstellen und nicht eine stetige Funktion; die stets endliche Anzahl verhindert die völlige Anschmiegung an die Wahrscheinlichkeitskurve und kann immer nur eine Annäherung bedeuten. Und wenn wir die Verteilung in der anderen Form, entsprechend dem Schema der Wahscheinlichkeitsmaschine betrachten: wir hatten gesehen, daß die Verwirklichung eines solchen Schemas die Unabhängigkeit zweier Vorgänge voraussetzt, und daß diese stets nur mit Näherung erreicht werden kann. Die völlige Unabhängigkeit aber würde dem Kausalprinzip widersprechen und würde zwar logisch widerspruchsfrei, aber dennoch nicht vorstellbar sein. Es hat keinen Sinn, den Begriff eines zufälligen, als eines nichtkausalen, Geschehens zu bilden; denn vorstellbar sind nur Begriffe, deren Gegenstand – bei Allgemeinbegriffen nach Hinzukommen spezieller Bestimmungen – in der Erfahrung angetroffen werden kann.[24] Nur als von einem Grenzfall können wir von einem zufälligen Geschehen sprechen; wir können aussagen, daß es Geschehnisse gibt, deren kausaler Einfluß aufeinander so klein ist, daß wir diesen vernachlässigen können, und die Grenze, der eine solche geringe Abhängigkeit zustrebt, können wir Zufall nennen. Dennoch ist der Grenzfall selbst nicht als wirklicher Vorgang denkbar. Man kann den Begriff des Zufalls dem Unendlichkeitsbegriff vergleichen, der auch nur die bloße Negation eines vorstellbaren Begriffs bedeutet. Das Seltsame ist nun, daß die Wahrscheinlichkeitsurteile gerade von solchen Grenzfällen Aussagen machen, und deshalb es prinzipiell ausgeschlossen ist, daß ihr Gegenstand jemals in der Erfahrung angetroffen wird. Stets ist ihr Gegenstand nur mit Näherung verwirklicht, aber diese Näherung ist anders als bei den geometrischen Sätzen, die nur wegen der Unvollkommenheit unseres Erkenntnisvermögens Näherungen bleiben müssen.

24. Vgl. Nators Äußerung, die sich zwar nur auf logischen Widerspruch im Begriff bezieht, aber auch für den Widerspruch zu den synthetischen Grundsätzen des Verstandes Geltung besitzt. „Schließlich ist die logische wie jede Terminologie bis zu gewissem Grade der Willkür anheimgegeben; aber zweckmäßig ist es wohl nicht, Begriff zu nennen, wodurch nicht etwas begriffen wird. Ich verstehe unter Begriff eine vollziehbare Denkeinheit; der Widerspruch aber ist im Denken unvollziehbar. Die Aussage, daß etwas ein Widerspruch sei, setzt nicht den Begriff eines Widersprechenden voraus, sondern nur die Frage, ob aus gegebenen begrifflichen Elementen eine Denkeinheit vollziehbar sei oder nicht.“ (S. 102).

objects that rarely occur in nature but that can be applied to nature as an approximation. However, the ideal object is *imaginable* and it is not ruled out that a real object might at some point assume exactly the shape of one of the simpler of these ideal objects. We even have to assert that every object, even if we never completely know its form, has at every time instant a particular spatial shape and that we, if we only knew this shape, could assert particular laws about it with certainty. The ideal objects of geometry are realized in every instant even though we do not know them in any detail. In the case of the theorems of probability, the situation is different. There is no event that constitutes the ideal case of the probability calculus. No matter how often I repeat a process, | the values 74
of the frequencies will form a discrete sequence and not a continuous function; since the number is always finite, hugging the probability curve can only ever be an approximation. And the distribution in case of the schema of the probability machine showed us that the realization of such a process requires the independence of two processes and that this can only ever be accomplished approximately. Complete independence would contradict the principle of causality and even though it would be consistent, it would still not be imaginable. It makes no sense to consider the concept of coincidental but causally unrelated events, since the only concepts that are imaginable are those whose content – in the case of general concepts after addition of some special determination – can be experienced.[24] We can only speak of coincidental events as a limiting case; we can say that there are events whose causal influence on each other is so small that we can neglect it, and the limit to which such a small dependence tends we may call coincidence. Nevertheless, the limiting case itself cannot be imagined as a real process. One can compare the concept of coincidence to that of infinity which is also just the negation of an imaginable concept. The peculiar situation is that judgments of probability make claims about such limiting cases and therefore it is in principle impossible that their content can ever occur in experience. Their content is only ever realized as an approximation, but this approximation is different than that for geometrical theorems which must remain approximate because of the incompleteness of what we can know.

24. See Natorp's comment. Although it only refers to the concept of logical contradiction, it is also valid for the synthetic principles of the intellect. "Ultimately the logical, as any other terminology, is subject to a certain degree of arbitrariness; but it does not seem useful to call something that cannot be grasped a concept. I take a concept to be a completable unit of thought; but a contradiction is not a completable thought. The claim that something is a contradiction does not presume the concept of contradiction, but only the question whether the given elements of concepts form a completable unit of thought or not." (p. 102).

75 Trotzdem bleiben die Wahrscheinlichkeitsgesetze notwendige Bestandteile unseres Wissens, und es ist ein Irrtum, zu glauben, daß mit wachsender physikalischer Erkenntnis der Wahrscheinlichkeit eine immer geringere Rolle in der mathematischen Darstellung der Wirklichkeit zufallen werde. Zwar werden die Lücken unserer Kenntnis in der fortschreitenden Entwicklung ausgefüllt werden, aber die festen Regeln über das Geschehen in der Natur, wie sie in speziellen Wahrscheinlichkeitssätzen der Physik niedergelegt sind, werden darum nicht geändert werden. Wenn wir einmal in der Lage sein sollten, genau zu berechnen, welche Seite eines bestimmten Würfels nach dem 30. Wurf oben liegen wird, so wird damit nichts an der Tatsache geändert sein, daß das Häufigkeitsverhältnis für alle Seiten nahezu gleich sein wird. Vom Standpunkt der Nützlichkeit erscheint sogar die Aussage des Wahrscheinlichkeitsgesetzes oft vorteilhafter als die der speziellen Berechnung. Denn der einzelne Vorgang wird häufig gleichgültig sein gegenüber dem Gesamtresultat der ganzen Reihe; dies kann wichtige physikalische Gesetzmäßigkeiten enthalten, während der Einfluß des einzelnen Vorgangs auf das wirkliche Geschehen belanglos ist. In sehr klarer und bestimmter Weise hat Cournot den Laplaceschen Gedanken, daß eine vollkommene menschliche Intelligenz keine Wahrscheinlichkeitsgesetze mehr benützen würde, abgetan: «Une intelligence supérieure à l'homme ne différerait pas de l'homme qu'en ce qu'elle se tromperait moins souvent que lui, ou même ne se tromperait jamais dans l'application de cette donnée de la raison. Elle ne serait pas exposée à regarder comme indépendantes, des séries qui s'influencent réellement, dans l'ordre de la causalité, ou inversement, à se figurer une dépendance entre des causes réellement indépendantes. Elle ferait avec plus grande sûreté, ou même avec une exactitude rigoureuse, la part qui revient au hasard dans le développement des phénomènes successifs. Elle assignerait a priori les résultats du concours de causes indépendantes, ce que nous sommes le plus souvent dans l'impuissance de faire. Par exemple, étant donné un dé de structure irrégulière, qui doit être projeté un grand nombre de fois, par des forces impulsives dont l'intensité, la direction, et le point d'application sont déterminés à chaque coup par des causes indépendantes de celles qui agissent aux coups suivants, elle saurait, ce que nous ne savons pas, quel doit être à très peu près le
76 | rapport du nombre des coups amenant une face déterminée, au nombre total des coups; et cette science aurait un objet certain, soit qu'elle connût les forces qui agissent et qu'elle en pût calculer les effets, pour chaque coup particulier, soit que cette connaissance et ce calcul surpassassent[E4] is a encore ses forces. En un mot, elle pousserait plus loin[E5] que nous et appliquerait mieux la science de ces rapports mathématiques, tous liés à la notion du hasard, et qui deviennent des lois de la nature, dans l'ordre des phénomènes».[25]

25. Cournot, S. 82 [Reichenbach erroneously cites p. 104]

E4. Reichenbach incorrectly copied the tense from Cournot. Reichenbach's original read "surpassent".
E5. Reichenbach's original erroneously read "loins" instead of "loin" here.

Nevertheless the laws of probability remain necessary parts of our know- 75
ledge and it would be a mistake to believe that probability will become less important for the mathematical representation of the world as physical knowledge grows. Of course, the gaps in our understanding will be filled by continuous development, but the fixed rules for natural events that physics provides in the special theorems of probability will not be changed thereby. Should we ever be able to calculate which side will come up on the 30th throw of a die, that will change nothing about the fact that the ratio of frequencies will be roughly the same for all sides. In terms of its utility, the claim made by the law of probability often seems more advantageous than the calculation of a specific case. Since the particular process will often be indifferent for the overall result of the whole sequence, which may contain important physical regularities, while the influence of the individual process on the real event is irrelevant. Cournot dismissed in a very clear and concise way the Laplacian thought that a complete human intelligence would have no use for the laws of probability:

> "An intelligence superior to man would differ from man only in making fewer mistakes, or in making no mistakes at all in the application of this idea of reason.[E11] [This intelligence] would not be tempted to consider series that influence each other in the order of causality as independent [from each other], or, conversely, it would not imagine dependence among independent causes. It would determine with greater precision, or even with rigorous exactitude, the role of chance in a sequence of phenomena. It would establish a priori the results of the combination of independent causes, which, in most cases, we cannot do. For example, given an irregular die that is to be thrown many times according to impulsive forces whose intensity, direction, and point of application are determined for each throw by causes independent of the forces that will affect the throws that follow it, the superior intelligence would know – something that
> we do not know – very precisely, the ratio | of the number of throws 76
> bringing up a designated face of the die to the total number of throws; and this science [of probabilities] would have a definite content, [no matter] whether [the superior intelligence] has knowledge of the active forces and the possibility of calculating their effects for each particular throw, or if this knowledge and this calculation still are beyond its powers. In a word, the superior intelligence would go beyond us and would better apply the science of these mathematical relations that are connected to the notion of chance, and that become laws of nature, within the order of phenomena."[25]

25. Cournot, p. 82. [Reichenbach erroneously cites p. 104.]

E11. Cournot is referring here to a thought he mentions in the previous paragraph: the combination of many independent causes and the analysis of sequences of independent trials.

Cournots Worte, die er vor 70 Jahren niederschrieb, sind durch ein neuerdings bekannt gewordenes Beispiel aufs beste illustriert worden. Es gibt in der Tat einen Fall, wo wir jedes einzelne spezielle Resultat genau berechnen können, und wo wir doch in der Lage sind, Wahrscheinlichkeitsgesetze anzuwenden; dies ist das Poincarésche Logarithmenproblem. Wenn wir die dritten Dezimalen einer Logarithmentafel auszählen, so ist nach den Gesetzen der Wahrscheinlichkeit anzunehmen, daß ebenso viele gerade wie ungerade Ziffern angetroffen werden; trotzdem wir jeden einzelnen Logarithmus genau berechnen können, ja diese Berechnung bereits ausgeführt vorliegt, muß die Wahrscheinlichkeitsbetrachtung sich auf die Verteilung noch anwenden lassen. Poincaré hat mathematisch bewiesen, daß in der Tat die Gleichverteilung auf gerade und ungerade Ziffern mit beliebiger Genauigkeit stattfindet, wenn man die Intervalle des Numerus und die Dezimale entsprechend wählt.[26] Es ist dies ein besonders instruktives Beispiel dafür, daß die Wahrscheinlichkeit nicht, wie es die Theorie der subjektiven Wahrscheinlichkeit vertritt, einen Ausweg der menschlichen Unwissenheit darstellt; denn hier, wo jede Unwissenheit ausgeschaltet und jeder einzelne Fall genau berechnet ist, gelten dennoch dieselben, objektiven Wahrscheinlichkeitsgesetze.

Die theoretische Physik ist heute dazu geschritten, Wahrscheinlichkeitsbetrachtungen in umfangreichem Maße zur Darstellung des wirklichen Geschehens heranzuziehen; und eines ihrer wichtigsten Gesetze, das Entropieprinzip, beruht auf solchen Voraussetzungen. Alle derartigen statistischen Betrachtun-
77 gen beziehen sich nicht auf die Wiederholung des Vorgangs in der Zeit, son|dern auf seine Vervielfältigung im Raume; daß auch für diese das Prinzip der Wahrscheinlichkeitsfunktion Gültigkeit besitzt, ist im früheren gleichfalls gezeigt worden. Es ist jedoch noch keine exakte Analyse der Maxwell-Boltzmannschen Statistik gegeben worden, die die zugrunde liegenden philosophischen Prinzipien klar genug aufdeckt; und wenn auch die flüchtige Betrachtung lehrt, daß ein anderes als das hier deduzierte Prinzip der Wahrscheinlichkeitsfunktion nicht darin auftritt, so muß doch die Antwort der Philosophie auf die Probleme der physikalischen Statistik bis zur Durchführung einer solchen Untersuchung aufgeschoben werden.

26. Poincaré, *Probabilité*, p. 313.

Cournot's words, which he wrote 70 years ago, are best illustrated by a recent example. There is in fact a case where we are able to calculate each individual specific result exactly and where we are nevertheless able to apply laws of probability; namely Poincaré's problem of logarithms. If we count the third decimals of a logarithm table, then according to the laws of probability we can assume that there will be the same number of even and odd digits, even if we can compute the logarithm exactly. In fact, if we already have the calculation in front of us, the approach using probability is still applicable to the distribution. Poincaré proved mathematically that the equal distribution of even and odd numbers does in fact occur with arbitrary precision, if one chooses the intervals for the whole number and decimal appropriately.[26] This is a particular instructive example that shows that probability is not, as the theory of subjective probability holds, an escape for human ignorance; since here, where all ignorance has been removed and every individual case has been computed exactly, the same objective laws of probability still hold.

Nowadays, theoretical physics uses probabilistic considerations extensively to represent real events; and one of its most important laws, the principle of entropy, is based on such assumptions. All such statistical considerations refer, not
to the repetition of the process in time, | but rather to many copies in space. It 77
has also been shown above that the principle of the probability function holds for these, too. However, no sufficiently clear and exact analysis of the philosophical basis of the Maxwell-Boltzmann statistic has been given, and even if a superficial scan shows that it contains no other than the principle of the probability function deduced here, an answer from philosophy to the problem of physical statistics will nevertheless have to be postponed until the completion of such an analysis.

26. Poincaré, *Probabilité*, p. 313.

## 79 Lebenslauf

Ich, Hans Friedrich Herbert Günther Reichenbach, evangelisch-reformierter Konfession, Hamburgischer Staatsangehörigkeit, wurde am 26. September 1891 als Sohn des Kaufmanns Bruno Reichenbach und seiner Ehefrau Selma Reichenbach, geb. Menzel, zu Hamburg geboren. Ich besuchte von Ostern 1898 bis Ostern 1907 die Realschule Stiftungsschule von 1815 zu Hamburg, an der ich das Berechtigungszeugnis zum einjährig-freiwilligen Dienst erhielt, von Ostern 1907 bis Ostern 1910 die Oberrealschule vor dem Holstentore zu Hamburg, an der ich die Abiturientenprüfung bestand. Ich wandte mich dann dem Studium der Bauingenieurwissenschaft zu und bezog die Technische Hochschule zu Stuttgart, an der ich von Ostern 1910 bis Ostern 1911 immatrikuliert war. Bald jedoch vertauschte ich das Ingenieurstudium mit dem der theoretischen Wissenschaften und ging an die Universität Berlin. Ich studierte Philosophie, Mathematik, Physik, Pädagogik: in Berlin von Ostern 1911 bis Ostern 1912, in München von Ostern 1912 bis Ostern 1913, in Berlin von Ostern 1913 bis Ostern 1914. Seit Ostern 1914 bin ich in Göttingen immatrikuliert. Ich hörte Vorlesungen und Praktika u. a. bei folgenden Herren:

In Stuttgart: Prof. Faber, Hammer, Mehmcke, Sauer.

In Berlin: Prof. Cassirer, Frobenius, Knoblauch, Planck, Riehl, Rubens, Dr. Rupp, Prof. Simmel, Struve, Stumpf.

In München: Prof. v. Aster, Dr. Fischer, Prof. Pringsheim, Sommerfeld.

In Göttingen: Prof. Bernstein, Debye, Hilbert, Husserl.

## Resume 79

I, Hans Friedrich Herbert Günther Reichenbach, of reformed protestant denomination and of Hamburgian nationality, was born in Hamburg, September 26, 1891, as son of the merchant Bruno Reichenbach and his wife Selma Reichenbach, née Menzel. I attended the middle school Stiftungsschule von 1815 in Hamburg from Easter 1898 to Easter 1907, where I received the certificate authorizing a one year voluntary service. From Easter 1907 to Easter 1910 I attended the high school Oberrealschule vor dem Holstentore in Hamburg, where I passed my Abitur exams.[E12] I began studying engineering at the Technische Hochschule in Stuttgart, where I was enrolled from Easter 1910 to Easter 1911. I soon decided to switch from engineering to the study of the theoretical sciences and went to the University of Berlin. I studied philosophy, mathematics, physics and pedagogy: in Berlin from Easter 1911 to Easter 1912, in Munich from Easter 1912 to Easter 1913 and in Berlin from Easter 1913 to Easter 1914. Since Easter 1914 I have been enrolled in Göttingen. I attended lectures and labs by, among others, the following gentlemen:

In Stuttgart: Prof. Faber, Hammer, Mehmcke and Sauer.

In Berlin: Prof. Cassirer, Frobenius, Knoblauch, Planck, Riehl, Rubens, Dr. Rupp, Prof. Simmel, Struve, Stumpf.

In Munich: Prof. v. Aster, Dr. Fischer, Prof. Pringsheim, Sommerfeld.

In Göttingen: Prof. Bernstein, Debye, Hilbert, Husserl.

E12. German final high school exam that enables entry into university.

# Bibliography


Apelt, E. F. 1854. *Theorie der Induktion*. W. Engelmann.

Bernstein, F. 1907. "Über das Gaußsche Fehlergesetz" *Mathematische Annalen* 64, p. 417.

Cassirer, E. 1910. *Substanzbegriff und Funktionsbegriff*. B. Cassirer.

Cournot, M. A. A. 1843. *Théorie des Chances et des Probabilités*. Hachette.

Czuber, E. 1891. *Theorie der Beobachtungsfehler*. B. G. Teubner.

——. 1908. *Wahrscheinlichkeitsrechnung*. B. G. Teubner.

Elsas, A. 1889. "Kritische Betrachtungen über die Wahrscheinlichkeitsrechnung" *Philosophische Monatshefte*, p. 557.

Fick, A. 1883. *Philosophischer Versuch über die Wahrscheinlichkeiten*. Stahel.

Grelling, K. 1910. "Die philosophischen Grundlagen der Wahrscheinlichkeitsrechnung" *Abhandlungen der Friesschen Schule III* 3.

Hausdorff, F. 1901. "Beiträge zur Wahrscheinlichkeitsrechnung" *Bericht der mathematisch-physischen Klasse der Königlich Sächsischen Gesellschaft der Wissenschaften* 53, pp. 152–178.

Kant, I. 1787. *Kritik der reinen Vernunft*. 2. Aufl. Akademieausgabe.

Lange, F. A. 1894. *Logische Studien*. Baedeker.

Laplace, P. S. 1840. *Essai Philosophique sur les Probabilités*. Bachelier.

Natorp, P. 1910. *Die logischen Grundlagen der exakten Wissenschaften*. B. G. Teubner.

Poincaré, H. 1912. *Calcul des Probabilités*. Gauthier-Villars.

——. 1914. *Wissenschaft und Methode*. B. G. Teubner.

Stumpf, C. 1892a. "Über den Begriff der mathematischen Wahrscheinlichkeit" *Sitzungsbericht der philosophisch-historischen Klasse der königlich bayerischen Akademie der Wissenschaften zu München.*

——. 1892b. "Über die Anwendung des mathematischen Wahrscheinlichkeitsbegriffs auf Teile eines Continuums" *Sitzungsbericht der philosophisch-historischen Klasse der königlich bayerischen Akademie der Wissenschaften zu München.*

von Kries, J. 1886. *Die Prinzipien der Wahrscheinlichkeitsrechnung.* J. C. B. Mohr.

# Index